AF812401

Green Energy and Technology

Climate change, environmental impact and the limited natural resources urge scientific research and novel technical solutions. The monograph series Green Energy and Technology serves as a publishing platform for scientific and technological approaches to "green"—i.e. environmentally friendly and sustainable—technologies. While a focus lies on energy and power supply, it also covers "green" solutions in industrial engineering and engineering design. Green Energy and Technology addresses researchers, advanced students, technical consultants as well as decision makers in industries and politics. Hence, the level of presentation spans from instructional to highly technical.

Indexed in Scopus.

More information about this series at http://www.springer.com/series/8059

Benjamin Duraković

PCM-Based Building Envelope Systems

Innovative Energy Solutions for Passive Design

Springer

Benjamin Duraković
Faculty of Engineering and Natural Sciences
International University of Sarajevo
Sarajevo, Bosnia and Herzegovina

ISSN 1865-3529 ISSN 1865-3537 (electronic)
Green Energy and Technology
ISBN 978-3-030-38334-3 ISBN 978-3-030-38335-0 (eBook)
https://doi.org/10.1007/978-3-030-38335-0

© Springer Nature Switzerland AG 2020
This work is subject to copyright. All rights are reserved by the Publisher, whether the whole or part of the material is concerned, specifically the rights of translation, reprinting, reuse of illustrations, recitation, broadcasting, reproduction on microfilms or in any other physical way, and transmission or information storage and retrieval, electronic adaptation, computer software, or by similar or dissimilar methodology now known or hereafter developed.
The use of general descriptive names, registered names, trademarks, service marks, etc. in this publication does not imply, even in the absence of a specific statement, that such names are exempt from the relevant protective laws and regulations and therefore free for general use.
The publisher, the authors and the editors are safe to assume that the advice and information in this book are believed to be true and accurate at the date of publication. Neither the publisher nor the authors or the editors give a warranty, expressed or implied, with respect to the material contained herein or for any errors or omissions that may have been made. The publisher remains neutral with regard to jurisdictional claims in published maps and institutional affiliations.

This Springer imprint is published by the registered company Springer Nature Switzerland AG
The registered company address is: Gewerbestrasse 11, 6330 Cham, Switzerland

Foreword

I feel very much delighted to write foreword for this new book titled *PCM-Based Building Envelope Systems: Innovative Energy Solutions for Passive Design*. My past experience has given me the opportunity to work in the area of phase change materials for various thermal applications. During this journey of my research, I have been able to benefit from the novel work of the author of this book "Benjamin Durakovic" on various aspects of phase change materials. Dr. Benjamin is an eminent faculty member of International University of Sarajevo. He is known for his research work that encompasses energy storage using phase change materials for passive building design, building energy efficiency, design of experiments, and data analysis. His tremendous contributions in more than 20 research and development projects in Bosnia and the USA, which lead to the publications of various important research papers in leading world-class journals in aforementioned fields, merit him as the best candidate to compile this book for the benefit of students, faculty members, engineers, researchers, and scientists working in building envelope systems around the globe.

This book is a ready reference to access details of energy storage passive design techniques and mathematical models using phase change materials. This book with eight unique chapters; in the opening chapter, first introduces the emerging problem of depleting resources and emphasizes on the significance of the building energy demands for heating and cooling purposes. Chapter 2 details about the various types of available phase change materials with their application limitations. Chapters 3–6 with pivotal significance provide details on the passive solar heating and cooling concepts. PCM-heat modulation with emphasis on thermal mass and free cooling is further discussed. The use of PCM directly in the building structures and components with possible methods is provided in Chaps. 4 and 5. Chapter 6 discusses the use of PCM as separate heat storage modules. In the later part of the book, the final chapter enlightens the reader with the physics of the processes involved during the heating and cooling of PCMs. Chapter 7 greatly contributes to discuss the heat transfer modes and mechanisms involved and finally elaborates the work on available simulation and modeling tools for PCM-based building envelope systems.

In a nutshell, this book will prove to be a complete guide for those already working in the domain of building envelope systems and looking for innovative energy solutions using phase change materials. For the beginners in the field of building envelope systems, this book will provide a complete walk through to the various stages and will help the reader to advance to the next level with detailed fundamental and applied knowledge.

October 2019

Dr. Hafiz Muhammad Ali
Associate Professor of Mechanical Engineering
King Fahd University of Petroleum and Minerals
Dhahran, Saudi Arabia
e-mail: Hafiz.ali@kfupm.edu.sa

Preface

Introduction

This book provides the latest research in the field of thermal energy storage technologies applicable for solar heating and cooling with the aim of reducing fossil fuel dependency. Particularly, the book discusses issues and advantages of common PCMs applicable for buildings as a more efficient novel solution for passive solar heating/cooling strategies. Up-to-date PCM-based energy storage solutions within building structure (wall, floor, ceiling, façade), components (windows, shading devices), and separate heat and cold storage devices are discussed in details. With the aim of building energy performance assessment, the book provides advanced modeling and simulation tools as theoretical base for the analysis of PCM-based building envelope in terms of heat storage and transfer.

Audience

This book will be of great value to those who deal with building energy analysis such as students, researchers, and professionals involved in the field of mechanical and civil engineering as well as architectural design.

The Aim of the Book

In the last few decades, world energy demand based on limited non-renewable sources was rapidly increased. The participation of the building sector through heating and cooling in global energy consumption is notable and takes up to 40%. To reduce building energy demand and carbon emission, it is necessary to implement innovative design concepts such as passive strategies. Heat storage plays one of the most important roles in building energy demand reduction. Emerging technologies based on the application of phase change materials (PCMs) in building envelope enhance the thermal performances of buildings and reduce energy demand.

Passive building design based on PCM is very active research area and new results and techniques become available almost on the continuous basis. The aim of this book is to present the state of the art in this research field including novel achievements based on the application of PCMs. Therefore, the book represents a collection of novel methods, techniques, and the recent research results including own research. Key problems and possible solutions were discussed.

Book Structure

In this book, *PCM-Based Building Envelope Systems: Innovative Energy Solutions for Passive Design*, passive design techniques based on phase change materials for energy storage as well as basics of the mathematical models were discussed. Each chapter has abstract, body, and conclusion. Graphical explanations and tables were used within the body to present key information in a quick way. Figure 1 represents graphical structure of the book.

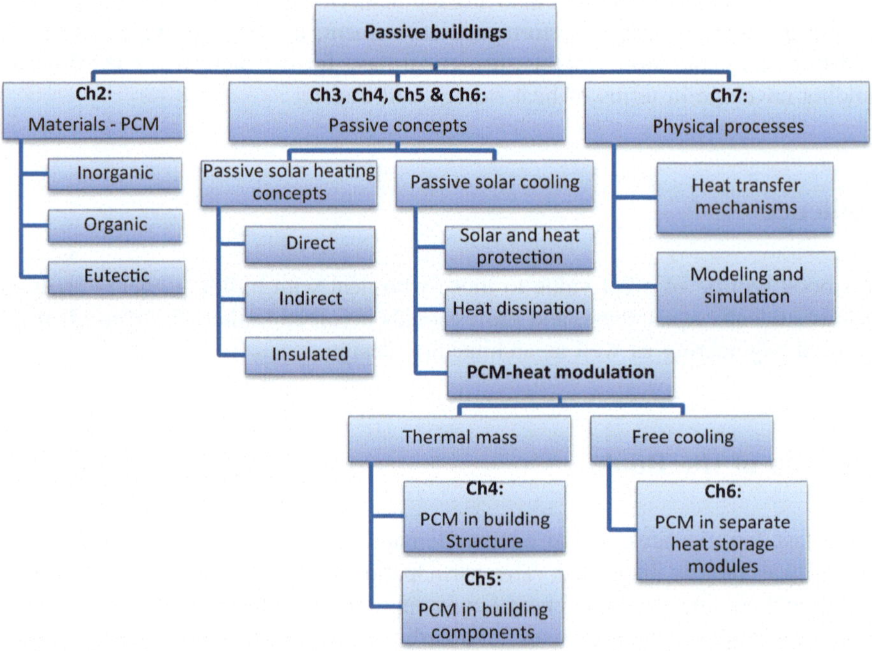

Fig. 1 Book structure

The book is organized into three major areas that are important for passive building design. In the first part, phase change materials and its properties, and advantages, disadvantages, and possibilities to be used for heat storage in passive building concepts were discussed in Chap. 2. In the second part, passive design techniques for heating and cooling were discussed. Particularly, the working principle of each technique was discussed with recent research developments and application. The part that deals with PCM-based energy storage techniques within the building structure was discussed in detail in Chap. 4; the PCM-based energy storage within building components was discussed in details in Chap. 5, and the PCM-based energy storage in separate heat storage modules was discussed in Chap. 6. Finally, the basics of the mathematical models for heat transfer modes and numerical simulation were introduced in Chap. 7.

Special attention was paid to the glazing systems as the weakest point of buildings in terms of heat gain/loss. Therefore, the aim in this case is to search for new technologies to reduce the energy consumption demand in buildings by considering different glazing system designs as responsive building elements and applying new materials. These systems contribute in reducing on-peak energy demand. Special focus is on the advantages of surplus heat gains made at midday stored in the thermal mass and its availability to the load offset later in the day when the outside temperature drops in winter/summer mode. In addition, to conduct comparative study of glazed systems performances for those based on the conventional technologies versus responsive glazed system technology with the aim of identifying the most appropriate system as a design solution.

Sarajevo, Bosnia

Benjamin Duraković, Ph.D.
e-mail: bdurakovic@ius.edu.ba

Acknowledgements Many people have contributed their time to improve the quality of the book. I express my sincere appreciation and gratitude to all of them. I express my sincere gratitude to my friends and colleagues who supported and encouraged me during this journey in challenging moments. I also wish to thank to various publishers and authors who have given permission to reproduce their material in this book.

Special thanks to Dr. Hafiz Muhammad Ali for taking his valuable time to read the manuscript and write the book foreword. Special recognition is due to my family who constantly has much patience with me and has contributed greatly toward success of this book.

About This Book

In the last few decades, world energy demand based on limited non-renewable sources was rapidly increased. The participation of the building sector through heating and cooling in global energy consumption is notable and takes up to 40%. To reduce building energy demand and carbon emission, it is necessary to implement innovative design concepts such as passive strategies. Heat storage plays one of the most important roles in building energy demand reduction. Emerging technologies based on the application of phase change materials (PCMs) in building envelope enhance the thermal performances of buildings and reduce energy demand. These materials use latent heat of fusion to store relatively large amounts of energy for later usage, at narrow temperature ranges.

This book provides the latest research in the field of thermal energy storage technologies applicable for solar heating and cooling with the aim of reducing fossil fuel dependency. Particularly, the book discusses issues and advantages of common PCMs applicable for buildings as a more efficient novel solution for passive solar heating/cooling strategies. Up-to-date PCM-based energy storage solutions within building structure (wall, floor, ceiling, façade), components (windows, shading devices), and separate heat and cold storage devices are discussed in detail. With the aim of assessment of building energy performance, the book provides advanced modeling and simulation tools as theoretical base for the analysis of PCM-based building envelope in terms of heat storage and transfer.

This book will be of great value to those who deal with building energy analysis such as students, researchers, and professionals involved in the field of mechanical and civil engineering as well as architectural design.

Benjamin Duraković

Contents

About the Author

Benjamin Duraković is Assistant Professor at Faculty of Engineering and Natural Sciences, International University of Sarajevo. His research interests are energy storage using phase change materials for passive building design, building energy efficiency, design of experiments, and data analysis. He has authored or co-authored many research papers in the aforementioned topics and published in a variety of leading journals including *Sustainable Cities and Society, International Journal of Low-Carbon Technologies, Journal of Materials and Environmental Science*. He teaches courses at both undergraduate and post-graduate levels in the above-mentioned areas. Apart from this, he has completed or participated in more than 20 R&D innovation projects with Bosnian and US companies. He is actively involved as reviewer for leading journals including *Energy conversion and management, Energy and Buildings, Construction and Building Materials*, and he is Journal Editor of *Periodicals of Engineering and Natural Sciences*. e-mail: bdurakovic@ius.edu.ba

Chapter 1
Introduction

Strong economic and population growth in the last three decades caused a rapid increase in the world energy consumption producing an enormous amount of greenhouse gases (GHG). Fossil fuels are primary sources of CO_2, which take a share of about 65% in total GHG. Nowadays, there is a global trend of reducing greenhouse gasses. This reduction is primarily based on research and investigation of different innovative technologies that would positively affect GHG reduction. As one of the main contributors to the production of greenhouses is buildings and construction sector through energy consumption, building energy demand reduction is being directed toward research and development of innovative technologies that would decrease fossil fuel consumption.

Today, energy conversion and electricity generation are mainly based on burning fossil fuels and nuclear power (fission or fusion), which all rely on finite resources generating waste end products in an irreversible cycle Fig. 1.1. Fossil and nuclear fuel reserves are limited and generate huge amounts of waste and pollutant emissions. At the current consumption rate, coal reserves will run out in about 130 years, natural gas in about 60 years, and oil in about 40 years [1].

Based on Fig. 1.1, about three-quarters of power generation comes from non-renewable sources, which have very limited reserves. Building energy takes about 40% of global energy demand. Applying passive design strategies can contribute to the reduction in building energy demand.

1.1 Depletion of the Non-renewable Energy Resources

Coal consumption has never stopped increasing and has been the world's fastest-growing energy source in recent years—faster than gas, oil, nuclear, hydro, and renewable energy as shown in Fig. 1.2 [3]. This trend will continue in the future where coal will be a key component of the energy mix. Total world coal production

© Springer Nature Switzerland AG 2020

B. Duraković, *PCM-Based Building Envelope Systems*, Green Energy and Technology, https://doi.org/10.1007/978-3-030-38335-0_1

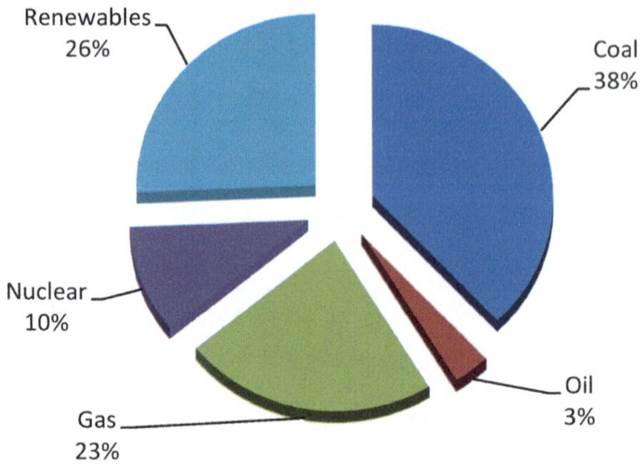

Fig. 1.1 World electricity production from all energy sources in 2018 [2]

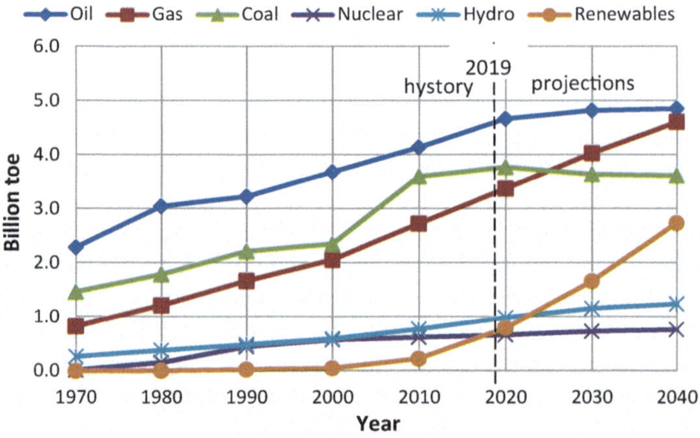

Fig. 1.2 World energy consumption by 2040 [2]

reached a record level in 2012, increasing by 2.9% in comparison with the previous year. Coal plays a key role in power generation and currently fuels 38% of the world's electricity. The figure is much higher in some individual countries. South Africa relied on coal for 94% of its electricity and Poland for 81% in 2014 [2] (Fig. 1.3).

In fact, it is expected that the share of the coal in electricity generation will be reduced below 30%, while the renewables will overtake gas and coal as the most used form of primary energy as of 2040.

World oil consumption is well reflecting the increase in the consumer demand for petroleum products, and it is on track to become critically low in 40 years. This means humankind cannot afford to wait 40 years, and must urgently conduct energy-saving programs and transition to alternative energy sources in the following

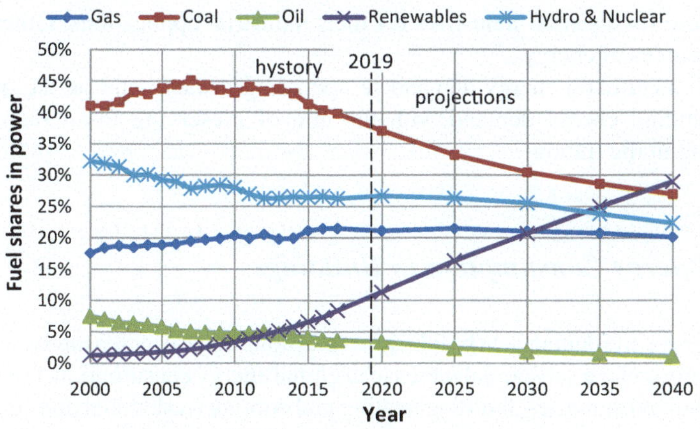

Fig. 1.3 Fuel shares in world electricity generation [2]

two decades. While this figure is hotly debated, what is clear is that oil has a host of useful industrial applications and to irreversibly burn oil endangers the future. World crude oil demand has been growing at an annualized compound rate slightly in excess of 2.0% per year recently. Demand growth is highest in the developing world, particularly in China and India. High demand growth is primarily caused due to rapidly rising consumer demand for transportation via cars and trucks powered with internal combustion engines. For economic and/or political reasons, this high demand growth component did not exist in most of the developing world even a decade ago [4].

Natural gas consumption is the fast-growing fossil fuel with estimated reserves for 60 years. Total world consumption of natural gas for industrial uses increases by an average of 1.5% per year through 2040, and consumption in the electric power sector grows by 2.0% per year. Growth in consumption occurs in countries outside of the Organization for Economic Co-operation and Development (OECD) member countries, where demand increases more than twice as fast as in OECD countries. Growth in natural gas consumption is particularly high in non-OECD countries, where economic growth leads to increased demand over the projection period. Consumption in non-OECD countries grows by an average of 2.2% per year through 2040, more than twice as fast as the 1.0% annual growth rate for natural gas demand in the OECD countries [5].

Oil, gas, and coal are a precious resource that humankind cannot afford to burn completely, thus consumption of these resources must be slowed down. Someday, currently unknown sources of fossil fuels may be discovered, but it would be irresponsible to base today's energy decisions on such an uncertain prediction. One is sure that continuation of oil burning can harm the viability of the future because oil is needed for lubricating the machines of the world for years to come, as well as to secure its continued use in the petrochemical industry [6]. Thus, we do need to sustain these fossil-based industries for their industrial applications, rather than as primary sources of energy.

How to meet extra energy demand in upcoming decades and on the other side reduce building energy demand with the aim of preserving fossil fuels will be challenging in the future.

1.1.1 Energy Consumption in Buildings

In recent years, the demands in building thermal comfort rise increasingly, which correspondingly cause a substantial increase in global energy consumption. For instance, in industrialized countries, building heating and cooling (residential and commercial energy consumption) participate between 20 and 40% in total energy consumption [7]. Windows can account for between 30 and 50% of the energy losses in buildings [4]. Opposed to other industrialized countries, building heating was accounted for 63% of natural gas consumed in US homes; the remaining 37% was for water heating, cooking, and miscellaneous uses [4]. In the European Union, the building sector is consuming 40% of the global energy, and two-thirds of this energy consumption is due to the heating, ventilation, and air-conditioning (HVAC) systems. Therefore, it is critical to reduce the energy demand of the buildings [8].

To reduce energy demand in buildings and contribute to preserving fossil fuels can be achieved by passive building design, using renewable energy sources, improving the thermal insulation of building envelope and storing energy for later usage. Solar energy is one of the most considered energy sources for building space heating, cooling, or solar hot water between 2010 and 2020, and generation from renewables grows 5.2% per year, compared with 3.9% per year between 2000 and 2010. Estimated global electricity generation from renewable energy sources will grow 2.7 times between 2010 and 2035 [9].

Solar heating systems use solar energy to heat the interior space or to store the energy in a storage system for later use. The heat is transferred by a heat transfer fluid (HTF), which can be either liquid or air. Liquid systems are more often used when storage is included and are well suited for radiant heating systems, boilers with hot water radiators, and even absorption heat pumps and coolers. Solar space cooling is an innovative technology that converts heat collected from the sun into useful cooling for applications such as building space conditioning. Obtained heat through the use of solar collectors is converted into cold using a "sorption" cooling process. The resulting cold is delivered to the application using an HTF (chilled water or dry cool air). Solar hot water is a simple and relatively inexpensive way to reduce energy consumption in residential and commercial buildings. The solar hot water system is comprised of solar collectors and an automatic control unit that work together. By harnessing the power of the sun, the systems can reduce the need for electricity 25–50%. In case that the solar system cannot provide adequate space heating, an auxiliary system provides the additional heat.

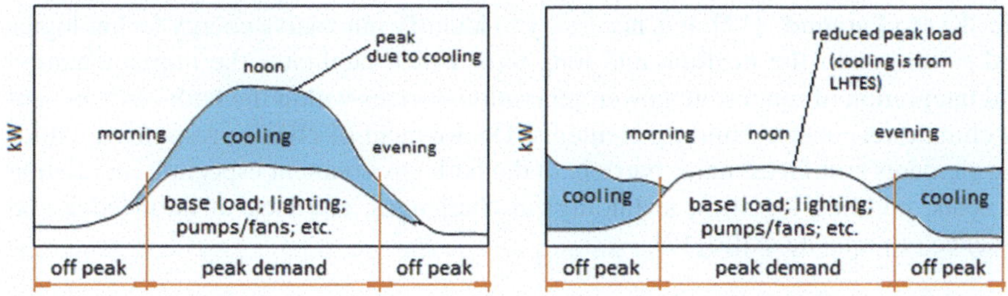

Fig. 1.4 Electrical profile—conventional building using chillers versus TES electrical profile

Different shading devices, solar attic cooling, and façade cooling can reduce energy demand in buildings as well. Hot attics cause the entire house to be warmer. Attic air temperatures can reach over 70 °C. Super-hot air gets trapped and collects in attics and causes heat to back up in the building.

Emerging technologies that improve building envelope performance have significant potential to reduce building energy consumption. Actual savings from these technologies will depend heavily upon their performance in diverse climate and operational conditions. Performance levels are determined to overcome existing window technologies as well. Overall heating and cooling savings depend mostly on how well solar heat is blocked [10].

Energy and its storage are one of the biggest technology challenges of the upcoming decades. Thermal energy storage (TES) systems as responsive building elements can contribute to reducing on-peak electricity demand. Latent heat thermal energy storage (LHTES) is an attractive technique for accumulating energy because it provides a high-energy storage density per unit of mass at constant or near-constant temperature. Thus, LHTES requires a smaller amount of phase change material (PCM) for the same amount of stored energy. The principle of PCM is a simple endothermic and exothermic reaction.

Referring to Fig. 1.4, the largest moveable portion of the building's electrical load can be shifted from high-cost "on-peak" hours to low-cost "off-peak" hours. TES/LHTS system is charged by capturing low-cost ("off-peak") electricity from the power grid during the night and then using this captured energy for cooling over the following day ("on-peak" hours). Another way is capturing solar heat over the day and using it over the night for heating in the wintertime.

As Fig. 1.4 shows, in conventional systems, the chiller must be run only when the building occupants want cool air. In a thermal energy storage system, the chiller can be run at times other than only when the occupants want cooling, as much as half of the energy consumed in a commercial building is used for cooling and heating. By using properly designed TES solutions, often up to 30–40% of this energy use can be saved [11]. The ability to store energy for use at peak demand is a challenge for delivering efficient innovative technologies and solutions.

Based on the recommendations given in IEA ECBCS (*International Energy Agency, Energy Conservation in Buildings & Community Systems*) Annex 44, and the “Kyoto Pyramid” [12], it is necessary to identify innovative energy technologies and solutions for the medium and long term which facilitates the implementation and integration of renewable power generation devices within the built environment as climate-responsive building elements. Deployment of climate-responsive building elements still faces major barriers in the built environment especially in relation to costs, building logistics, technological challenges, and lack of knowledge and absence of requisite skills [12].

1.2 Thermal Energy Storage

Thermal energy can be stored as sensible heat, latent heat, or as chemicals energy. Figure 1.5 shows the thermal energy storage forms with examples.

The scope of this book is to elaborate only latent heat thermal energy storage within phase change materials that have solid to liquid-phase transition. These materials belong to organic and inorganic groups of materials, which is discussed in detail in Chap. 2. Figure 1.6 shows comparisons of volumes of heat storage systems based on sensible heat, latent heat, and thermochemical heat storage for a passive house.

Figure 1.6 shows that thermochemical systems have the highest heat storage density but with significantly increased temperature, which is suitable for seasonal storage applications. The estimate is based on the energy demand of 6480 MJ for a passive house [13].

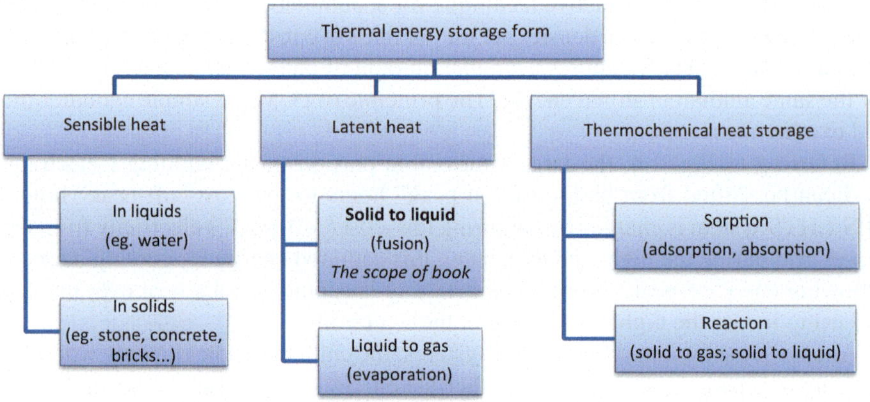

Fig. 1.5 Thermal energy storage

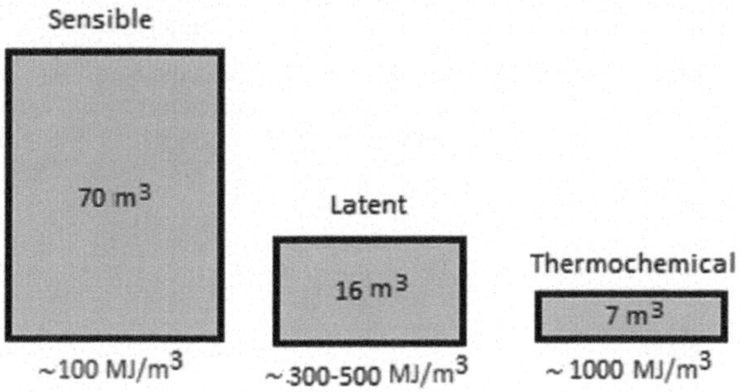

Fig. 1.6 Volume needed to fully cover the annual storage need of a passive house

1.2.1 Sensible Heat Storage

Sensible heat storage is based on the specific heat capacity of solids and liquids such as rocks, concrete, plasterboard, sand, water, and molten salts. The heat storage process is based on the temperature change of the storage medium. Sensible heat is the most common and the oldest way of energy storage that has been used for centuries in buildings. It is known as building thermal mass that increases thermal inertia of buildings, which contributes to temperature variance reduction and improves thermal comfort. Therefore, the storage capacity of the building envelope depends on the material used and composition of the wall. In this case, the building envelope will have better thermal storage performances if the insulation [14] is placed to the exterior wall layer than to the interior wall layer. Storage capacity (thermal mass) of building envelope may play a significant role in building energy consumption for heating. Heating energy reduction in a heavy weighted building may go up to 30% due to its increased thermal mass. Table 1.1 gives the summary of the properties of common construction material for sensible heat storage.

Construction materials are used for thermal energy storage within a building structure while liquids are used for heating or cooling. Water is one of the best materials as heat transfer fluid due to its availability increased specific heat capacity and very low cost.

1.2.2 Latent Heat Storage

Latent heat storage is based on enthalpy of solids and liquids. The advantage is based on a very narrow or constant range of temperature change during the energy storage/release. In most cases, solid to liquid-phase change process is used due to low volumetric expansion, where the fusion is used to store the heat and solidification is used to release the heat. Paraffins, fatty acids, salt hydrates, and eutectic mixtures

Table 1.1 Common construction material for sensible heat storage [15]

Material	Density (kg/m^3)	Thermal conductivity (W/mK)	Specific heat capacity (kJ/kgK)	Volumetric heat capacity (kJ/m^{3}K)
Gypsum (coating)	1000	0.4	1	1000
Gypsum (plasterboard)	900	0.25	1	900
Ceramic tile	2000	1	0.8	1600
Lime mortar	1600	0.8	1	1600
Cement mortar	1800	1	1	1800
Concrete	2000	1.35	1	2000
Concrete (high density)	2400	2	1	2400
Reinforced concrete (2%)	2400	2.5	1	2400
Cement bonded particleboard	1200	0.23	1.5	1800
Oriented strand board	600	0.14	1.7	1020
Oriented strand board	900	0.18	1.7	1530
Water (40 °C)	990	0.63	4.19	4148
Wood	450–700	0.12–0.18	1.6	720–1120
Plywood boards	500–1000	0.13–0.24	1.6	800–1600
Rock	2800–1500	3.5–0.85	1	2150
Limestone	1600–2600	0.85–2.3	1	2100
Sand and gravel	1700–2200	2	0.91–1.18	2072
Clay or silt	1200–1800	1.5	1.67–2.5	3252

are common representatives of PCMs for building applications. Table 1.2 shows the leading global suppliers of PCMs for a building application.

The temperature range of commercial PCMs is from −10 to +120 °C. The enthalpies of PCMs are ranging from 100 to 430 MJ/m^3. For comparison, sensible heat storage in water ranges from 50 to 250 for cooling and heating, respectively. Table 1.3 shows the comparative results between sensible heat and latent heat storage capacities.

For sensible heat storage materials, the energy is stored in the temperature range 25–75 °C.

Table 1.2 Leading supplier of PCMs for latent heat storage

Supplier	Location	Form of PCM	Type of PCM
TEAP	India	Bulk	Inorganic
Rubitherm GmbH	Germany	Bulk	Inorganic; organic
PureTemp	USA	Bulk	Organic
PlusICE	UK	Bulk	Inorganic; organic
Mitsubishi Chemicals	Japan	Bulk	Inorganic
Honey well	Germany	Bulk	Organic
Cristopia	France	Bulk	Inorganic
Climator	Sweden	Bulk	Inorganic
PCM Energy	India	Bulk;	Salt hydrates
Micronal	Germany	Bulk; microencapsulated	Organic, slurry
PCM Products Ltd	UK	Bulk; macroencapsulated	Sub zero eutectics, salt hydrates and organics
savENRG™	USA	Bulk; macroencapsulated	Organics and inorganic
BASF—Micronal PCM	Germany	Bulk; macroencapsulated	Powders

Table 1.3 Sensible heat storage versus latent heat storage [16]

Property	Sensible heat storage		Latent heat storage	
	Rock	Water	Paraffin wax	$CaCl_2 \cdot 6H_2O$
Density (kg/m^3) at 24 °C	2240	1000	1802	795
Specific heat capacity [kJ/(kg K)]	0.9	4.18	–	–
Enthalpy (kJ/kg)	–	–	174.4	266
Storage volume for storing 1 GJ (m^3)	9.9	4.8	3.2	4.7
Relative volume	3.1	1.5	1.0	1.5

1.2.3 Thermochemical Heat Storage

Thermochemical energy storage operates as sorption processes and chemical reactions. In sorption processes, energy is stored either through adsorption (physical bonding) or absorption (material dissolution). In chemical reactions, energy is stored as the heat of reaction of reversible reactions. Negligible heat losses and high-energy density are the main advantages of thermochemical reaction heat storage.

The research in this field is in an early stage, and there are no thermochemical heat storage commercial solutions for building applications [13]. Some problems associated with these materials are costs and low thermal conductivity. The research is mainly focused on long-term storage solutions for solar energy. Currently, there are no materials available for reliable deployment of this technology, even though it has a high-energy density and long-term storage capacity. Therefore, a lot of research effort has to be done to improve efficiency and optimize the system.

1.3 History of PCM Application in TES

Reducing building energy demand is challenging but achievable, for example through better insulation of building envelope using new materials and approaches. Emerging technologies based on the application of phase change materials for energy storage [17] can enhance the thermal performances of buildings [18] but still other issues are present such as PCM degradation due to thermal cycling [19]. These materials use latent heat of fusion to store relatively large amounts of energy for later usage, at narrow temperature ranges [20]. Therefore, to increase building thermal mass and building thermal storage potential especially for lightweight structures is an important aspect of building design. Therefore, PCMs may play a significant role in stabilizing the interior temperature fluctuation.

Application of PCMs in buildings for thermal storage is stared in the late 1940s. The pioneer in this was mechanical engineer Maria Talkes, an assistant at MIT's Department of Metallurgy. She designed a PCM-based heating system for the Dover Sun House built about 30 km west of Boston in 1948 [21]. Phase change material, sodium sulfate decahydrate (Glauber's salt), was used as PCM. The PCM caused corrosion on the containers while cracks and leakage of the PCM appeared [22].

The building is renovated and returned to a water-based heat storage system. After a failed experiment with the Dover Sun House, she continued working on the PCM-based heat storage systems and in 1954 claimed that the experiment was successful. In the late 1950s, many PCM-based passive houses were built, including the one outside of the USA, in Casablanca, Morocco in 1957, for solar cooling [22].

The Dover house PCM-based heating system technologies have been evolved significantly over the decades, especially in the 1980s and 1990s [23–26]. Today, PCMs are well familiar material in the building design with its application in active heating/cooling systems [27], building structure [28], wallboards, bricks [29], plaster [30], concrete [31], glazing systems [32–34] etc. The trend change in PCM technology research is shown in Fig. 1.7.

The total number of papers found in Scopus for key words[1] is 2473. About 82% of the publications come from predominantly hot climate areas. Figure 1.7b shows the most research-intensive countries in this field.

Emerging technologies based on integration phase change material (PCM) in glazing and shading solutions show the potential to enhance the thermal performance of buildings [18]. These technologies are focused on reducing interior temperature

[1]Obtained from Scopus for key words in TITLE-ABS-KEY (pcm AND phase AND change AND material AND building).

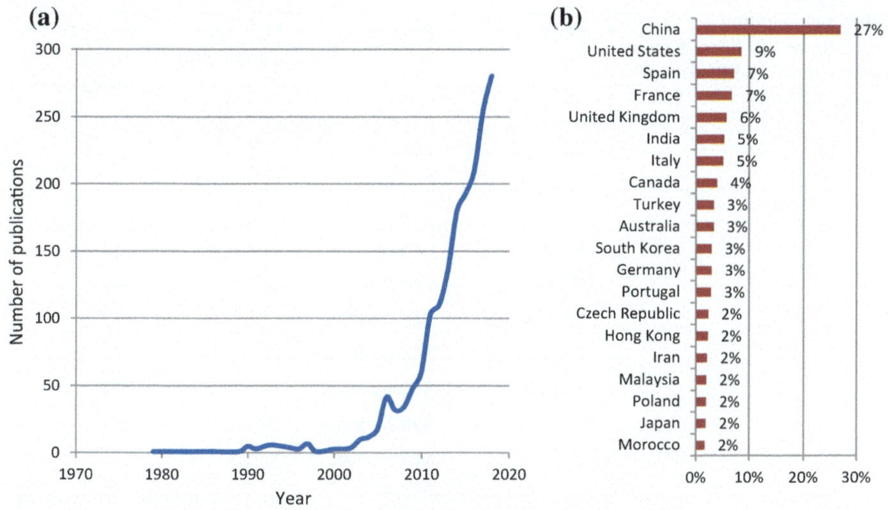

Fig. 1.7 Research trend change [Obtained from Scopus for key words in TITLE-ABS-KEY (pcm AND phase AND change AND material AND building)]

variation through storing solar energy in the form of latent heat for later usage. The application of these technologies in building components may provide a high-energy storage density per unit of mass at constant or nearly constant temperature [35, 36]. Heat gain to the interior can be reduced by extending the period of PCM melting, which can be provided by adding a thicker layer of PCM.

1.4 PCM-Based Technologies for Built Environment

PCMs have been tested as a thermal mass component in buildings for the past 50 years, and most studies have found that PCMs enhance building energy performance. Some problems, though, such as high initial cost, loss of phase change capability, corrosiveness (in cases of some inorganic PCMs), and PCM leaking have hampered widespread adoption. Paraffinic hydrocarbon PCMs generally perform well, but they increase the flammability of the building envelope [37]. For these reasons, more attention is paid to PCMs based on fatty acids or inorganic salt hydrates.

A general classification of phase change material in building application can be listed as the following:

- Passive heating/cooling strategies with PCM
- PCMs integrated into the building structure (wall, ceiling, and floor)—microencapsulation
- PCMs in building components (e.g., facade element, glazed systems)—macroencapsulation/microencapsulation

Table 1.4 Melting temperature and latent heat of some commercial PCMs for passive buildings

PCM	Product	Melting temperature (°C)	Heat of fusion (kJ/kg)
Astorstat HA17	Paraffins and waxes	21.7–22.8	–
Astorstat HA18	Paraffins and waxes	27.2–28.3	–
RT26	Paraffin	24–26	232
RT27	Paraffin	28	180
Climsel C23	Salt hydrate	23	148
Climsel C24	Salt hydrate	24	108
STL27	Salt hydrate	27	207
S27	Salt hydrate	29	188
–	Mixture of two salt hydrates	22–25	–
E23	Plus ICE	23	155

- Active heating/cooling strategies with PCM
- PCMs in separate heat and cold storage devices—macroencapsulation/microencapsulation.

Incorporating PCM in the building structure and components represents passive systems that automatically release stored heat when indoor or outdoor temperatures fall below the melting point of PCM. Only PCMs that have a phase transition temperature close to thermal comfort temperature can be used in building applications as a passive system. The temperature range for this purpose is 21–29 °C. Typical commercial PCMs for passive building applications are presented in Table 1.4.

Separate heat and cold storage devices require active components such as control systems, fans, and pumps to move the air and heat transfer fluid (HTF). The main advantage of this system is the accessibility to the stored heat when it is required. Various temperatures can be used depending on the area of application. Since the occupant comfort temperature requires a range between 20 and 27 °C depending on cooling/heating season, the storage temperatures of the PCM are favored in the range of 0 and 40 °C (for cooling/heating). The exception of that is hot water and heating water provision where the temperature range is of 50 and 60 °C.

PCM in building components is considered a glazed system for thermal energy storage applicable for summer and winter modes. Computer simulation and experimental method were applied [38]. Performances of the model were investigated for different cases of the glazed unit orientation and environmental conditions. From the exterior environment, the glazed system is exposed to solar radiation and convection. The part of the radiation is reflected in the exterior, absorbed by the glazing system, and transmitted to the interior. Absorbed component of the radiation is emitted partially to the exterior and partially to the interior in the form of longwave radiation.

Conduction through the glazing system exists from exterior to the interior and vice versa depending on the weather condition or seasons. The interior side of the glazing is exposed to the convection as well.

1.5 Remarks

Based on the foregoing, the significance of this book is reflected in the contribution to the development of climate-responsive building components based on the energy storage with the aim of reducing building energy demand and so preserving fossil fuels. This is also a contribution to the development and application of new technologies; therefore, it is a good reference for architects, engineers, researcher, academicians, and students.

Historical evolution of PCM-based latent heat energy storage systems and importance of its application in buildings are discussed. Thermal energy storage systems based on renewables are notable in reducing dependency on conventional energy sources and in contributing to a more economical way of energy use [39]. Applications of phase change materials in passive design techniques enhance the energy storage and the functionality of the system. Since PCMs have high thermal mass, it was a subject of interest for a number of researchers.

References

1. World Coal Association (2013) Coal statistics. World Coal Association [Online]. Available http://www.worldcoal.org/resources/coal-statistics/. Accessed 13 Feb 2014
2. IEA (2019) Global energy and CO_2 status report. International Energy Agency [Online]. Available https://www.iea.org/geco/data/. Accessed 9 Aug 2019
3. World Bank database (2014) The World Bank [Online]. Available http://data.worldbank.org/. Accessed 5 Nov 2015
4. U.S. Energy Information Administration (2014) Heating and cooling no longer majority of U.S. home energy use. 7 March 2013 [Online]. Available http://www.eia.gov/. Accessed 13 Feb 2014
5. U.S. Energy Information Administration (2013) International energy outlook 2013. EIA, Washington, D.C.
6. Abbott D (2010) Keeping the energy debate clean: how do we supply the world's energy needs? Proc IEEE 98(1):42–66
7. Pérez-Lombard L, Ortiz J, Pout C (2008) A review on buildings energy consumption information. Energy Build 40(3):394–398
8. Navarro L, de Garcia A, Solé C, Castell A, Cabeza LF (2012) Thermal loads inside buildings with phase change materials. Energy Procedia 30:342–349
9. International Energy Agency (2012) Renewable energy outlook—WEO2015. International Energy Agency
10. DeForest N, Shehabi A, Garcia G, Greenblatt J, Masanet E, Lee ES, Selkowitz S, Milliron DJ (2013) Regional performance targets for transparent near-infrared switching electrochromic window glazings. Build Environ 61:160–168

11. OzGreen Energy (2014) HVAC thermal energy storage systems [Online]. Available http://www.ozgreenenergy.com.au/services_1.html. Accessed 16 Feb 2014
12. IEA International Energy Agency (2011) Annex 44 integrating environmentally responsive elements in buildings. In: IEA international energy agency, energy conservation in buildings and community systems, 2011 [Online]. Available http://www.ecbcs.org/annexes/annex44.htm. Accessed 20 Feb 2014
13. Tatsidjodoung P, Pierrès NL, Luo L (2013) A review of potential materials for thermal energy storage in building applica. Renew Sustain Energy Rev 18:327–349
14. Durakovic B, Yildiz G, Yahia ME (2020) Comparative performance evaluation of conventional and renewable thermal insulation materials used in building envelops. Tehnicki vjesnik—Technical Gazette 27(1) (In Press)
15. ISO 10456:2007 (2007) ISO 10456:2007—Building materials and products—hydrothermal properties—tabulated design values and procedures for determining declared and design thermal values. ISO
16. Hasnain SM (1998) Review on sustainable thermal energy storage technologies, Part I: heat storage materials and techniques. Energy Convers Manag 39(11):1127–1138
17. Khyad A, Samrani H, Bargach MN (2016) State of the art review of thermal energy storage systems using PCM operating with small temperature differences: focus on Paraffin. J Mater Environ Sci 7(4):1184–1192
18. Silva T, Vicente R, Rodrigues F (2016) Literature review on the use of phase change materials in glazing and shading solutions. Renew Sustain Energy Rev 53:515–535
19. Goia F, Boccaleri E (2016) Physical–chemical properties evolution and thermal properties reliability of a paraffin wax under solar radiation exposure in a real-scale PCM window system. Energy Build 119:41–50
20. Benomar W, Zennouhi H, Arid A, Rhafiki TE, Msaad AA, Kousksou T (2015) PCM building material under cyclic melting and freezing processes. J Mater Environ Sci 6(12):3416–3422
21. Grant L (1948) Sun heated home ready for test through winter near Boston. New York Herald Tribune, pp 17–17
22. Barber DA (2016) A house in the sun: modern architecture and solar energy in the cold war. Oxford University Press, New York
23. Mancini NA (1980) Use of paraffins for thermal storage. In: Proceedings of international TNO symposium on thermal storage of solar energy, Amsterdam
24. Lane A (1980) Low temperature heat storage with phase change materials. Int J Ambient Energy 1:155–168
25. Salyer I, Sircar A (1989) Development of PCM wallboard for heating and cooling of residential buildings. In: Thermal Energy Storage Research Activities Review, New Orleans, Mar 1989, pp 15–17
26. Bromley A, McKay E (1994) Incorporating phase change materials into the building fabric. In: Proceedings of CIBSE conference
27. Torlak M, Delalić N, Duraković B, Gavranović H (2014) CFD-based assessment of thermal energy storage in phase-change materials (PCM). In: Energy technologies conference—ENTECH'14, Istanbul, Turkey, 22–24 Dec 2014
28. Soares N, Costa J, Gaspar A, Santos P (2013) Review of passive PCM latent heat thermal energy storage systems towards buildings' energy efficiency. Energy Build 59:82–103
29. Castell A, Martorell I, Medrano M, Prez G, Cabeza L (2010) Experimental study of using PCM in brick constructive solutions for passive cooling. Energy Build 42(4):534–540
30. Liu H, Awbi H (2009) Performance of phase change material boards under natural convection. Build Environ 44(9):1788–1793
31. Cabeza L, Castellón C, Nogus M, Medrano M, Leppers R, Zubillaga O (2007) Use of microencapsulated PCM in concrete walls for energy savings. Energy Build 39(2):113–119
32. Duraković B, Mešetović S (2019) Thermal performances of glazed energy storage systems with various storage materials: an experimental study. Sustain Cities Soc 45:422–430
33. Duraković B, Torlak M (2017) Experimental and numerical study of a PCM window model as a thermal energy storage unit. Int J Low-Carbon Technol 12(3):272–280

34. Duraković B, Torlak M (2017) Simulation and experimental validation of phase change material and water used as heat storage medium in window applications. J Mater Environ Sci 8(5): 1837–1846. ISSN: 2028-2508 Copyright © 2017
35. Whiffen T, Riffat SB (2013) A review of PCM technology for thermal energy storage in the built environment: Part I. Int J Low-Carbon Tech 8(3):147–158
36. Whiffen T, Riffat SB (2013) A review of PCM technology for thermal energy storage in the built environment: Part II. Int J Low-Carbon Tech 8(3):159–164
37. Kissock J, Kelly J, Hannig M, Thomas I (1998) Testing and simulation of phase change wall-board for thermal storage in buildings. In: Proceedings of international solar energy conference, Albuquerque, New Mexico, 14–17 June 1998
38. Durakovic B (2017) Design of experiments application, concepts, examples: state of the art. Period Eng Nat Sci 5(3):421–439
39. Dinçer İ, Rosen MA (2002) Thermal energy storage—systems and applications. Wiley

Chapter 2
Phase Change Materials for Building Envelope

2.1 Introduction

The principle of PCM is endothermic and exothermic process. As the temperature of PCM increases, the PCM absorbs heat and changes phase from solid to liquid. Contrary, as the temperature of the PCM decreases, the PCM changes phase from liquid to solid and the PCM desorbs heat. The feasibility of using PCM in the latent heat storage system is based on desirable thermo-physical, kinetic, and chemical properties in addition to economic criteria [1]. The development of a latent heat thermal energy storage system requires the understanding of heat transfer in the PCMs when they undergo solid-to-liquid phase transition (and vice versa) in the required operating temperature range, as well as the design of the PCM container and formulation of the phase change problem [1].

Telkes and Raymond [2] were pioneers in phase change material research. Although it started in the 1940s, it did not take much attention until the energy crisis of the late 1970s and early 1980s. During the crisis, PCMs were extensively researched for use in different applications, especially for solar heating systems. Enormous work has been carried out to explore PCM usage in solar heating systems [3–6], as well as in other applications such as air-conditioning, building envelope [7–10], underfloor heating system [11, 12], electronics cooling [13–15], preservation of food, milk [16], waste heat recovery [17], textiles [18], etc. [19].

To be applicable in various applications, phase change materials must be encapsulated in a container (heat exchanger), to assure safe phase transition from solid to liquid and vice versa. Encapsulation means enclosing the material with an inert coating, which is required to prevent the external environment from contaminating the PCM. Two main approaches are used in encapsulation [20]:

- Macroencapsulation—PCMs are encapsulated in large bags, tubes, rectangular panels, or spherical capsules; and
- Microencapsulation—microscopic amounts of PCMs are coated with inert protective shell material.

© Springer Nature Switzerland AG 2020

B. Duraković, *PCM-Based Building Envelope Systems*, Green Energy and Technology, https://doi.org/10.1007/978-3-030-38335-0_2

Due to increased surface area, microencapsulation provides improved heat transfer between the PCM and its surroundings but usually adds cost because it involves more operations. Therefore, this section of the study reviews the papers on phase change materials for different applications published and reported in the scientific databases. Based on the analyzed papers, the major disadvantages of many PCMs are low thermal conductivities especially for the organic materials, i.e., low charging and discharging rates. Widespread usage of latent heat storage is limited due to the insufficient long-term stability of storage materials: poor thermal stability of the PCM and/or corrosion between the PCM and the container [19, 20]. The review covers usage of the PCM in different applications such as various designs of heat exchangers for energy storage including solar energy storage, waste heat recovery systems, and room heating and cooling.

2.1.1 Phase Change Theory

A diagram that shows different phases of a system under equilibrium is called a phase diagram. These diagrams represent the relationships between temperature and compositions and quantities of phases at equilibrium. For practical use, it is enough to consider only solid and liquid phases assuming the pressure to be constant at 1 atm. These diagrams do not indicate the dynamics when one phase transforms into another. However, it depicts information related to the microstructure and phase structure of a particular system. A phase diagram is actually a collection of solubility limit curves. The phase fields in equilibrium diagrams depend on the particular systems being depicted. For practical use, the PCM can be a pure substance, a eutectic mixture or a non-eutectic/binary mixture. The difference between eutectic and non-eutectic mixture is the phase change temperature. Eutectic PCM mixture is a special case of binary PCM mixture that has phase changes at a *constant* and *lowest temperature* compared to binary mixtures. Non-eutectic/binary PCM mixture has the phase changes in a *temperature interval* [21, 22].

Energy storage may be in the form of sensible heat (in a solid/liquid phase), latent heat, or as chemical energy, Fig. 2.1a. The phase change between the solid phase of

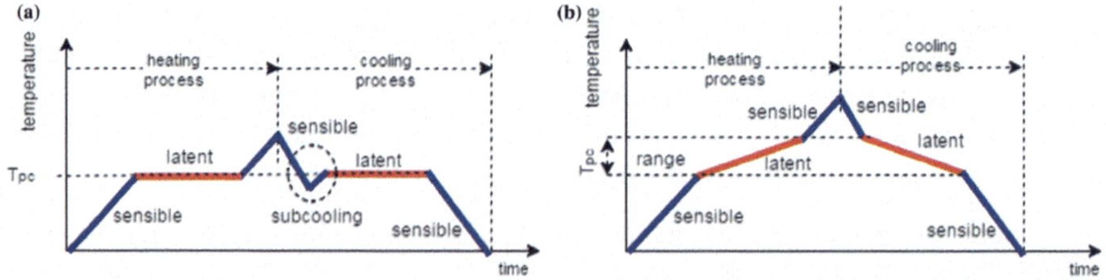

Fig. 2.1 Melting/solidification process: **a** eutectic with subcooling, **b** non–eutectic with temperature range

the material and its liquid phase is used to store latent heat energy. The transformation from the solid to liquid is called *melting* or *fusion,* and transformation from the liquid to solid is called *solidification*. The temperature at which the melting of the PCM begins is called *solidus temperature* and the temperature at which the liquid phase of the PCM begins to freeze is called *liquidus temperature* [22].

Referring to Fig. 2.1a, melting and cooling curves representing a pure/eutectic PCM are depicted. Sometimes, the liquid may cool to a temperature below its freezing point before crystallization occurs and this is called *supercooling* or *subcooling*, which is very important when dealing with the pure/eutectic PCM. This usually occurs when the heat is quickly removed from the liquid phase and the molecules have no enough time to form nucleation "seeds" that may trigger crystallization. If the liquid is undercooled, a bump on the solid–liquid interface can grow rapidly. The latent heat of fusion is removed by raising the temperature of the liquid back to the freezing temperature. A liquid below its standard freezing point will crystallize in the presence of a seed crystal or nucleus around which a crystal structure can form creating a solid [22].

Referring to Fig. 2.1b, a multi-component solidification of the PCM is depicted. In that case, most of the PCMs will solidify from the molten state over a range of temperatures. Thus, the cooling curve will have a liquid–solid transition between two different temperatures representing the beginning and end of solidification [22].

Therefore, the amount of stored energy as sensible heat can be expressed mathematically as follows:

$$Q = \int_{T_1}^{T_2} mC_p \mathrm{d}T = mC_p(T_2 - T_1) \tag{2.1}$$

Energy storage within PCMs is stored in the form of sensible heat in solid and liquid phases as well as latent heat during phase transition. The amount of stored energy can be expressed mathematically as:

$$Q = \int_{T_1}^{T_{\mathrm{pc}}} mC_p \mathrm{d}T + ma_{\mathrm{pc}}\Delta h_{\mathrm{pc}} + \int_{T_{\mathrm{pc}}}^{T_2} mC_p \,\mathrm{d}T \tag{2.2}$$

$$Q = mC_p\big(T_{\mathrm{pc}} - T_1\big) + ma_{\mathrm{pc}}\Delta h_{\mathrm{pc}} + mC_p\big(T_{\mathrm{pc}} - T_1\big) \tag{2.3}$$

where, m is mass in kg, C_p specific heat capacity of the body in J/kgK, T_1 and T_2 initial temperatures in solid state and final temperature in liquid state of PCM in K, T_{pc} phase transition temperature of PCM in K, Δh_{pc} enthalpy of phase change material in J/kg, and a_{pc} is a melted PCM fraction.

In Eq. (2.2), three components are presented. The first component represents sensible heat storage within solid phase before the temperature of the PCM reached phase change temperature (T_{pc}), the second one is the latent heat of phase change, and the third component represents the sensible heat storage within solid phase of the PCM after the melting process is completed.

From the literature review, most of the phase change materials used in practical purposes for heat storage have phase change temperature in the range of 20–60 °C [6].

2.1.2 PCMs for Energy Storage, Properties, and Classification

Most of the studies on storage materials have focused on sensible and latent heat storage systems. Using a latent heat storage system, a significant reduction in storage volume can be achieved using PCM compared to sensible heat storage. Thus, the latent heat storage system can be considered as a better way of storing thermal energy because of its high storage density. Sensible heat storage materials require typically 5–10 times higher storage mass than PCMs. For example, some study results show that rock as sensible heat storage material requires more than seven times storage mass than paraffin 116 wax (P116-Wax) [23–25]. Latent heat storage can be used in a wide temperature range depending on the application. A large number of PCMs are known to melt with a heat of fusion in any required range. The PCM to be used in the design of thermal storage systems should accomplish desirable thermo-physical, kinetics, and chemical properties given in Fig. 2.2 [26, 27].

Latent heat storage can be achieved through phase transformations such as solid–liquid, liquid–gas, solid–gas, and solid–solid. Only solid–liquid and solid–solid transformations are of practical interest. Phase transformations such as: solid–gas and liquid–gas have large volume changes on phase transition and, therefore, are less considered as thermal storage systems because it is impractical due to large changes in volume, although they have higher latent heat.

Figure 2.3 shows PCMs with solid–liquid phase transformation. Several authors have divided PCM into organic, inorganic, and eutectics PCMs [23].

Organic phase change materials are divided into two subgroups called paraffins and non-paraffins. Commonly known paraffin as organic material is wax. The non-paraffin organics are the most numerous of the phase change materials with highly varied properties. Abhat et al. [28] have conducted an extensive survey and recognized a number of esters, fatty acids, alcohols, and glycols suitable for thermal energy storage. These organic materials are subdivided into fatty acids and other non-paraffin organics [23]. They are flammable and should not be exposed to excessive high temperature, flames or oxidizing agents.

Inorganic phase change materials are divided in two subgroups called salt hydrates and metallics. Salt hydrates consist of salt and water that combine in a crystalline matrix when the material solidifies. There are many different salt hydrates having melting temperature ranges between 15 and 117 °C. Salt hydrates are considered as the most important group of PCMs that have been studied for application in latent

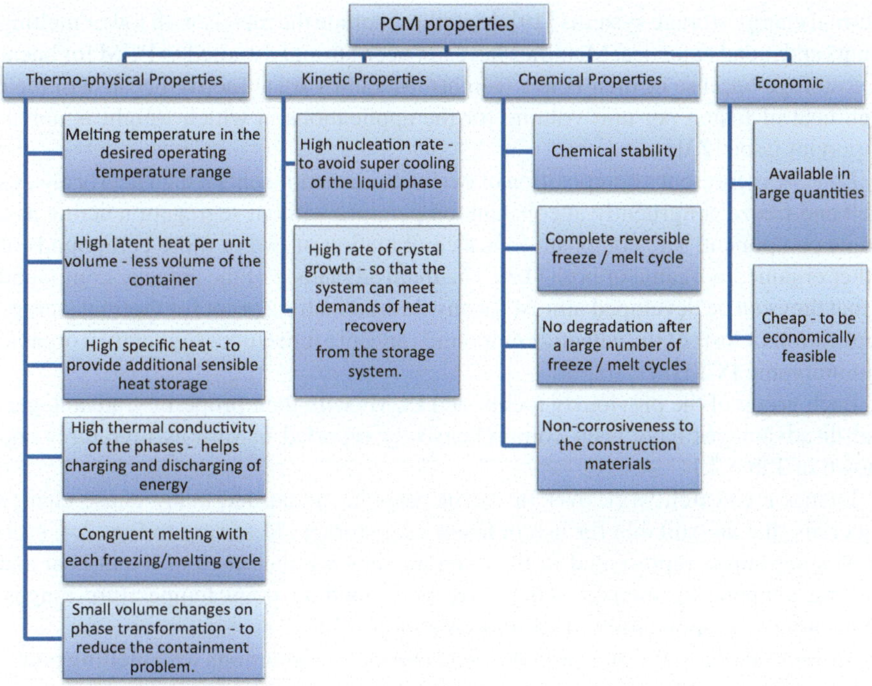

Fig. 2.2 Required properties for selection of a PCM

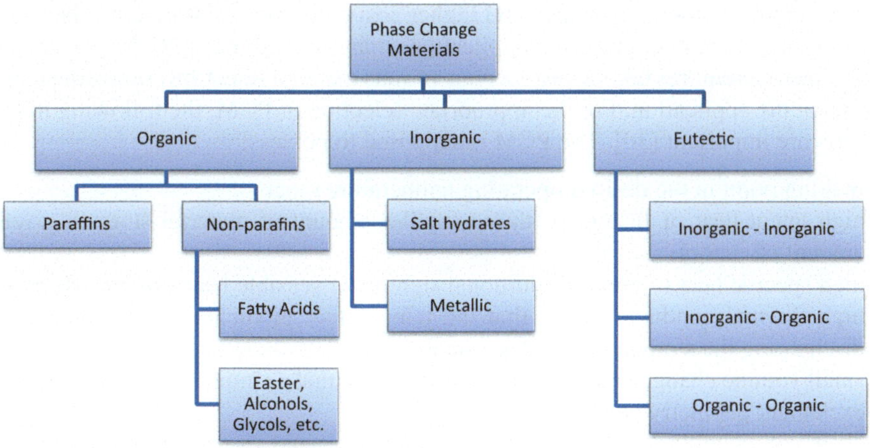

Fig. 2.3 Classification of phase change materials

thermal energy storage systems [29]. Metallics include the metals with lower melting points and metal eutectics. Metallics have not been strongly studied as PCM for latent heat storage because of their heavy weights. Metallics are attractive because of their high heat of fusion per unit volume for the applications in which weight is not an important issue [23].

Eutectics represent a composition of two or more components which nearly always melt and freeze congruently at constant temperature without segregation acting as a single component. The composition is a combination of two or more compounds of either organic, inorganic or both [19]. The main advantage of the eutectic compound is that they can be developed almost for any desired melting point for thermal storage applications. Costs of the eutectic compounds are upto three times higher than organic and inorganic PCMs.

Each group of the previously mentioned PCMs with their properties, advantages, and disadvantages have been comprehensively reported in various literatures and shown in Table 2.1.

Intensive research work over in recent years has identified many phase change materials that are suitable for use in latent heat storage. Referring to Fig. 2.4, each PCM subgroup is represented in the diagram showing the range of latent heat and melting temperature interval. They have wide melting point temperature ranges, which allow the application of various conditions [22].

Various mixtures of water and salts yield eutectic salt solutions with melting points significantly below 0 °C, which have numerous applications in the areas of heating, cooling, and air-conditioning. These materials may have high storage densities and are relatively inexpensive. Contrary, paraffins and fatty acids as main organic materials have lower storage densities and higher costs relative to salt hydrates but it is easier to use them in engineering applications than salt hydrates [22]. Selecting the right material requires taking into account various material selection criteria depending upon the application. For the appropriate selection of PCM, the following main criteria are important [30]. The PCM is supposed to possess:

- melting point in the desired operating temperature range
- high latent heat of fusion, so that a smaller amount of material stores a given amount of energy
- high specific heat to provide additional significant sensible heat storage effects
- high thermal conductivity, so that the temperature gradients for charging and discharging the storage material are small
- small volume changes during phase transition, so that a simple container and heat exchanger geometry can be used
- chemical stability, no chemical decomposition, and corrosion resistance to construction materials
- non-poisonous, non-flammable, and non-explosive elements/compounds, low cost for large quantities.

There is no single material that has all the required properties for an ideal thermal storage media; one has to select such PCM which will give as much performance as

Table 2.1 Properties of the PCM compared for each group [19]

	Organic materials		Inorganic materials		Eutectics
	Paraffins	Non-paraffins	Salt hydrates	Metallic	
Formula	C_nH_{2n+2}	$CH_3(CH_2)_{2n}{\cdot}COOH$	$AB^a{\cdot}mH_2O$	–	–
Melting point (°C)	−12 to 71	7.8–187	11–120	30–96	4–93
Heat of fusion	180–260 kJ/kg	130–250 kJ/kg	100–200 kJ/kg	25–90 kJ/kg	100–230 kJ/kg
Features	1. Melting point and latent heat increases with chain length 2. It is most used commercial PCM	1. Stearic acid melts over a wide range of temperature and has a large variation in latent heat of fusion	1. Oldest and most studied 2. Its alloys of inorganic salts and water	1. Not seriously consider, due to weight penalty	1. Composition of two or more components 2. Melts and freeze without segregation
Cost	Expensive	Two to three times costly than paraffins	Low cost	Costly	Costly
Advantages	1. No tendency of crystals to segregate and to supercool 2. Chemically stable 3. High heat of fusion 4. Compatible with all metal containers	1. Sharper phase transformation (direct transformation from solid to liquid without softening)	1. Easy availability 2. Sharpe melting point 3. High thermal conductivity 4. Low volume change than the others 5. Higher density 6. Non-flammable	1. High heat of fusion per unit volume 2. High conductivity 3. Non-flammable	1. High heat of fusion per unit volume 2. High conductivity

(continued)

Table 2.1 (continued)

	Organic materials		Inorganic materials		Eutectics
	Paraffins	Non-paraffins	Salt hydrates	Metallic	
Disadvantages	1. Low thermal conductivity. 2. Do not have sharp well-defined melting point 3. Flammable 4. High volume change	1. Mildly corrosive 2. Flammable, should not be exposed to excessively high temperature, flames or oxidizing agents	1. Supercooling 2. Corrosion on metal container 3. High volume change	1. Low heat of fusion per unit weight 2. Low specific heat	1. Low heat of fusion per unit weight
Examples Melting pt. (°C)/Latent heat (kJ/g)	*n*-tridecane (4.5/231), paraffin wax (32/251), *n*-tricontane (65/252)	Acetic acid (16.7/187), stearic acid (61/200), lauric acid (42/178), glycerin (18/198.7), bee wax (61.8/177)	$CaCl_2 \cdot 6H_2O$ (30/170–192), $Na_2SO_4 \cdot 10H_2O$ (32/251), $NaCl \cdot Na_2SO_4 \cdot 6H_2O$ (18/286), $MgSO_4 \cdot 7H_2O$ (48.4/200)	Gallium (30/80.3), gallium–gallium (29.8/-), Bi–Pb–In eutectic (70/29)	Na_2SO_4 + NaCl+ KCl + H_2O (4 /234), NH_2CONH_2 + NH_4NO_3 (46 /95)

[a]Represents non-water component in salt hydrate

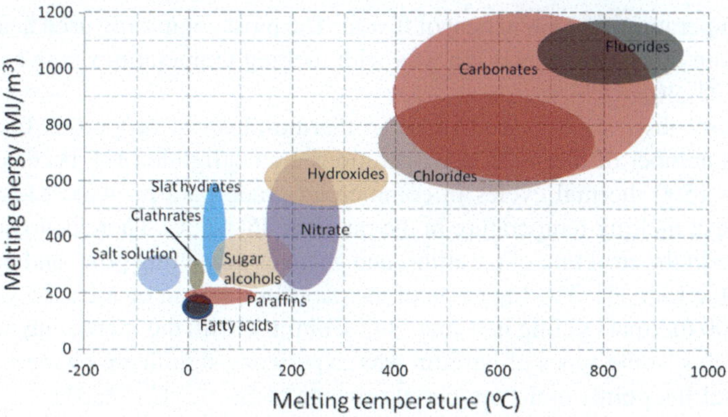

Fig. 2.4 Material classes that are being investigated and used as PCMs

possible at minimum cost. Selection of the PCM specifically for building applications depends on the following set of criteria [30]:

- region climate,
- electricity price,
- thermal cycling - stability,
- storing density (heat of fusion),
- costs,
- payback period.

Cost, stability, and *latent heat* are three key parameters that will decide the market adoption of PCMs. Paraffins are the most widely used PCMs in thermal energy storage applications because they are nontoxic, abundant in supply, easy to microencapsulate, have small subcooling, chemical inertness, and good recyclability. Paraffins are derived from crude oil and have sensitive prices to the season and to geopolitical scenarios. These reasons underline the need to shift the focus away from paraffins to another class of PCM. Bio-based PCMs and salt hydrates are two alternative groups of PCMs with great potential to substitute paraffins in the future. The selection of an optimal material for an engineering application from among two or more alternative materials on the basis of two or more attributes or criteria is a multiple attribute decision-making problem, which is out of the scope of this study [19].

2.1.2.1 Organic Phase Change Materials

Organic PCMs are classified as paraffins and non-paraffins and they can be melted and frozen repeatedly without phase segregation and without corrosiveness. From this group, paraffins have been used for thermal energy storage due to their high heat of fusion, varied phase change temperature, and non-toxicity. The paraffins of type C_nH_{2n+2} have similar properties, i.e., higher the value of n, the higher is the melting temperature, and latent heat of fusion. The most commonly used heat storage material from this group is paraffin wax with melting temperatures between 23 and 67 °C [20, 21, 30].

Laboratory tests show that changes in thermo-physical properties for paraffin wax after a number of thermal cycle operations are negligible [31]. Performed tests were up to 1500 thermal cycles in commercial grade paraffin wax. Most studied paraffins have melting temperature in the range of 45–62 °C. There is no noticeable degradation in the structure of paraffins, and changes in melting point and latent heat are negligible [32, 33]. The life cycle of the paraffin wax can be atleast 5 years with assumed 300 thermal cycling in a year [34]. Contrary, with the increasing number of thermal cycles, some types of paraffin wax experience significant changes in latent heat and melting points in the temperature range of 58–62 °C [33, 35].

Advantages of paraffin waxes

- They are chemically stable and show no tendency to segregate. They have slow oxidation when exposed to oxygen; therefore, they required closed containers [36].
- They did not show regular degradation in thermal properties after repeated melting/freezing cycles [32, 33].
- They have high enthalpies and have no tendencies to super cool, so nucleating agents are not necessary [34].
- They are safe and non-reactive [37].
- They are compatible with all metal containers and easily incorporated into heat storage systems [36].
- Attention should be paid when using plastic containers as paraffins have a tendency to infiltrate and soften some plastics [36] (Table 2.2).

Disadvantages of paraffin waxes

- They have low thermal conductivity in their solid state. This presents a problem when high heat transfer rates are required during the freezing cycle. This problem can be decreased using finned containers and metallic fillers, or through a combination of latent/sensible storage systems. For example, aluminum honeycombs have been found to improve system performance [34].
- They have a high volume change between the solid and liquid stages. This causes many problems in container design [37].
- They greatly decrease heat storage capacity. Unlike salt hydrates, commercial paraffins generally do not have sharp, well-defined melting points [36].
- They are flammable, but this can be easily solved by using a proper container [37].

Non-paraffins include fatty acids and other non-paraffin organics like esters, alcohol, glycols, etc. Fatty acids are most promising PCMs because of their suitable phase change temperature and high heat of fusion easily producible from common vegetable and animal oils [30, 36]. Fatty acids, characterized by the chemical formula $CH_3(CH_2)_{2n}COOH$, have much the same characteristics as paraffins. They are mildly corrosive but their advantage of sharper phase transformations is offset by the disadvantage of being about two or three times the cost of paraffins [37] (Table 2.3).

Table 2.2 Thermal cycling of paraffins

PCM	Melting point (°C)	Latent heat (kJ/kg)	Thermal cycles
Paraffins			
Paraffin (70 wt%) + Polypropylene (30 wt%)	44.77	136.16	3000
Paraffin ($C_{22.2}H_{44.1}$) (technical grade)	47.1	166	900
Paraffin ($C_{23.2}H_{48.4}$) (technical grade)	57.1	220	900
Paraffin wax 53 (commercial grade)	53	184	300
Paraffin wax 53 (commercial grade)	53	184	1500
Paraffin wax 54	53.32	184.48	1500
Paraffin wax 58–60	58.27	129.8	600
Paraffin wax 60–62	57.78	129.7	600
n-Heptadecane/polymethyl methacrylate ($C_{17}H_{36}$)	18.4	84.7	5000

Table 2.3 Thermal cycling of non-parafins

PCM	Melting point (°C)	Latent heat (kJ/kg)	Thermal cycles
Non-paraffins			
Acetamide (CH_3CONH_2)	82	263	1500
Acetamide (CH_3CONH_2)	82.15	262.78	300
Acetanilide (C_8H_9NO)	113	169.4	500
Capricacid (55 wt%) + expanded perlite (45 wt%)	31.80	98.12	5000
Erythritol	117	339	1000
Lauricacid ($C_{11}H_{23}COOH$)	43.5	169.3	120
Methylpalmitate	29	215	50
Methylstearate	37.8	270	50
Myristic acid ($C_{13}H_{27}COOH$)	50.4	189.4	450
Myristic acid ($C_{13}H_{27}COOH$)	52.99	181.0	1200
Myristic acid ($C_{13}H_{27}COOH$)	53.8	192.0	910
Palmitic acid ($C_{15}H_{31}COOH$)	61.2	196.1	120
Palmitic acid (80 wt%) + expanded graphite (20 wt%)	60.88	148.36	3000
Stearic acid ($C_{17}H_{35}COOH$)	65.2	209.9	450
Urea	133	250	50

The most common fatty acids for thermal storage are stearic acid, palmitic acid, lauric acid, and myristic acid. During freezing palmitic and lauric acids, small amount of supercooling is observed, which hamper their potential of use as heat storage [30]. Commercial grade stearic acid is thermally stable after repeated number of cycles [34]. Industrial-grade acids with 95% purity, (myristic, palmitic, and stearic with transition temperature 50–54, 58–62, and 65–69 °C), have at least 10% volumetric expansion when heated from room temperature to 80 °C and loose up to 10% of storage capacity after 450 thermal cycles with life cycle approximately a year [38]. Industrial-grade fatty acids with 90–95% purity, (stearic, palmitic, myristic, and lauric acids with melting points 53.8, 59.9, 53.8, and 42.6 °C). Industrial-grade PCMs have a tendency to change their thermal behavior. Palmitic acids (m.p. 59.9 °C) and myristic acids (m.p. 53.8 °C) may be considered as suitable PCMs in the long-term solar thermal applications. Stainless steel, carbon steel, aluminum, and copper are compatible with these acids, i.e., they are corrosion resistant [39].

2.1.2.2 Inorganic Phase Change Materials

Inorganic PCMs include salt hydrates and metallic PCMs. Salt hydrates are attractive for heat storage applications because of their low cost and easy availability [36]. They are the most studied PCMs in the form of pure substance or eutectic mixture in 1980s and 1990s [30, 36], with a sharp melting point and higher thermal conductivity than others [20, 26], and a high heat of fusion, which decreases the needed size of the storage system. They show a lower volume change than other PCMs, which makes it easy to design a container to accommodate volume change [34] (Table 2.4).

Table 2.4 Thermal cycling of salt hydrates

PCM	Melting point (°C)	Latent heat (kJ/kg)	Thermal cycles
Salt hydrates			
Calcium chloride hexahydrate ($CaCl_2 \cdot 6H_2O$)	29.8	190.8	1000
Glauber's salt ($Na_2SO_4 \cdot 10H_2O$)	32.4	238	320
Magnesium chloride hexahydrate ($MgCl_2 \cdot 6H_2O$)	111.5	155.11	500
$Na_2SO_4 \cdot nH_2O$	–	–	1000
$Na_2SO_4 \cdot 1/2NaCl \cdot 10H_2O$	20	–	5650
$NaOH \cdot 3 \cdot 5H_2O$	15	–	5650
Sodium acetate trihydrate ($NaCH_3COO \cdot 3H_2O$)	58	230	500
Trichlorofluoromethane heptadecahydrate ($CCl_3F \cdot 17H_2O$)	8.5	210	100

As main disadvantage of salt hydrates is phase separation and supercooling, so they have limited application. Other hydrates or dehydrated salts can be formed during a melting process. Some of salt hydrates (such as $Na_2SO_4 \cdot 10H_2$) decrease heat of fusion over 73% after 1000 thermal cycles. This problem can be eliminated by using gelled or thickened mixtures [34].

2.1.2.3 Eutectic Phase Change Materials

Investigations carried out on some of the eutectics whose phase change temperatures are suitable for low-temperature energy storage unit show that the thermal cycling can go upto 5000 cycles. The latent heat of fusion of eutectics is undergoing a phase transition within the temperature range 0–100 °C. In the thermal cycling of the organic eutectic for solar space heating applications, at the end of 1100 thermal cycles, the enthalpy change value was found to be 5% lower than its initial value. The thermal cycling of most inorganic eutectics shows good stability even after 1000 thermal cycles and shown no phase separation [19] (Table 2.5).

It is observed that most of the organic and inorganic eutectics which are proposed as PCMs are made from fatty acids and salt hydrates, respectively. The use of these materials as PCMs is very new to thermal storage systems, so there is very limited data available on thermo-physical properties including thermal cycling.

2.2 Properties of Commonly Used PCMs

PCM melting temperature (melting point and range), specific heat, heat fusion (enthalpy), thermal conductivity, and density are the most important characteristics of all PCMs [40]. These properties of some commonly available materials in the market were analyzed in order to find its mutual relationship and identify the best suited PCMs for building envelope integration as part of passive design.

2.2.1 Heat of Fusion Versus Melting Temperature

The heat of fusion is an important property of the PCMs. Higher heat of fusion is a desirable property of PCMs applied in passive design. The relation between the heat of fusion and melting temperature for selected common commercial PCMs is shown in Fig. 2.5.

Figure 2.5 represents the relation between heat of fusion (enthalpy) and melting temperature for some of the most common materials used as PCMs for building applications. The range between 0 and 18 °C represents the PCMs suitable for *active cooling* [41], such as: organic materials (Paraffin RT18, Butyl stearate–palmitate, Glycerin, Caprylic acid, Paraffin FMC, Rubitherm RT18 HC, Rubitherm RT21,

Table 2.5 Thermal cycling of eutectics

PCM organic eutectics	Melting point (°C)	Latent heat (kJ/kg)	Thermal cycles
Ammonium alum ($NH_4Al(SO_4)_2{\cdot}12H_2O$) (15%) + ammonium nitrate (NH_4NO_3) (85%)	53	170	1100
Butyl stearate (49wt%) + Butyl palmitate (48 wt%) + other (3 wt%)	17	138	100
Capric acid (65 mol%) + lauric acid (35 mol%)	13	116.76	120
Capric acid (73.5 wt%) + myristic acid (26.5 wt%)	21.4	152	5000
Capric acid (83 wt%) + stearic acid (17 wt%)	24.68	178.64	5000
Caprylic acid (70 wt%) + 1-dodecanol (30 wt%)	6.52	171.06	120
Lauric acid (66 wt%) + myristic acid (34 wt%)	34.2	166.8	1460
Lauric acid (69 wt%) + palmitic acid (31 wt%)	35.2	166.3	1460
Lauric acid (75.5 wt%) + stearic acid (24.5 wt%) +	37	182.7	360
Lauric acid (77.05 wt%) + palmitic acid (22.95 wt%)	33.09	150.6	100
Methyl stearate (86 wt%) + methyl palmitate (14 wt%)	23.9	220	50
Methyl stearate (91 wt%) + cetyl palmitate (9 wt%)	28.2	189	50
Methyl stearate (91 wt%) + cetyl stearate (9 wt%)	22.2	180	50
Myristic acid (58 wt%) + palmitic acid (42 wt%)	42.6	169.7	360
Myristic acid (64 wt%) + stearic acid (36 wt%)	44.1	182.4	1460
Myristic acid + glycerol	31.96	154.3	1000
Palmitic acid (64.2 wt%) + stearic acid (35.8 wt%)	52.3	181.7	360
Palmitic acid + glycerol	58.50	185.9	1000
Stearic acid + glycerol	63.45	149.4	1000
Inorganic eutectics			
$CaCl_2{\cdot}6H_2O$ (80 mol%) + $CaBr_2{\cdot}6H_2O$ (20 mol%)	20	117	1000

(continued)

Table 2.5 (continued)

PCM organic eutectics	Melting point (°C)	Latent heat (kJ/kg)	Thermal cycles
$CaCl_2{\cdot}6H_2O$(93wt%) + $Ca(NO_3)_2{\cdot}4H_2O$(5wt%) + $Mg(NO_3)_2{\cdot}6H_2O$ (2 wt%)	24	125	1000
$CaCl_2{\cdot}6H_2O$ (96 wt%) + KNO_3 (2 wt%) + KBr (2wt%)	23	138	1000
$CaCl_2{\cdot}6H_2O$ (96 wt%) + NH_4NO_3 (2 wt%) + NH_4Br(2wt%)	20	141	1000
$NaCH_3COO{\cdot}3H_2O$ (90 wt%) + $NaBr{\cdot}2H_2O$ (10wt%)	51	175	1000
$NaCH_3COO{\cdot}3H_2O$ (85 wt%) + $NaHCOO{\cdot}3H_2O$ (15wt%)	49	170	1000
$Mg(NO_3)_2{\cdot}6H_2O$ (93 wt%) + $MgCl_2{\cdot}6H_2O$ (7wt%)	78	152.4	1000

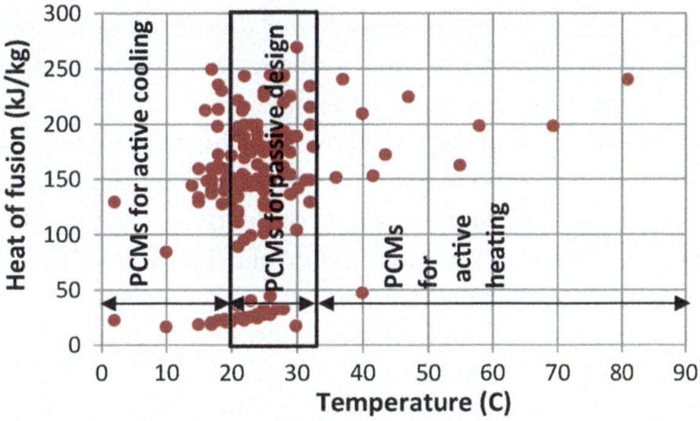

Fig. 2.5 Heat of fusion versus melting temperature

Rubitherm PX15), inorganic materials (Hexadecane, Heptadecane), and eutectics (PC14, PC17, PCM Products Ltd. S15, PCM Products Ltd. S17, PCM Products Ltd. A15, PCM Products Ltd. A16, PCM Products Ltd. A17, Microtek Laboratories MPCM18, Microtek Laboratories MPCM 18D, Emerest 2325). The range between 20 and 32 °C represents the PCMs suitable for *passive designing* [41], such as: organic materials (Paraffin C13–C24, Paraffin C17, Micronal DS 5001, Paraffin RT 27, almost all Paraffins, Salca, Rubitherms, Capric acid, Myristic acid, Palmitic acid, Stearic acid, and Acetamide), inorganic materials (*n*-Heptadecane, PlusICE PCM Hydrated salts, Climsel Climator Hydrated salts), and eutectics: (P116 Paraffin Wax, LA). The range between 33 and 100 °C represents the PCMs suitable for active heating [41], such as: eutectics: (CA-PA-SA) and inorganic materials: (Sodium sulfate decahydrate, Sodium thiosulfate pentahydrate).

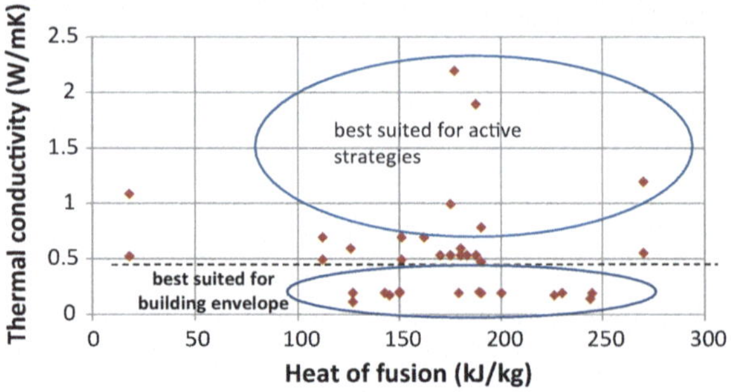

Fig. 2.6 Heat of fusion versus thermal conductivity for selected commercial PCMs

2.2.2 Heat of Fusion Versus Thermal Conductivity

Thermal conductivities of the PCMs are other important properties. Generally, a higher thermal conductivity is desirable if a PCM is used in heat exchangers such as active heating/cooling application. If the PCMs are integrated within the building envelope (which is the case with passive design), it is expected to act as an insulator [42]. Therefore, in this case, lower thermal conductivities of the PCMs are desirable. Figure 2.6 shows the ranges of thermal conductivities of PCMs for passive design with respect to the heat of fusion.

Figure 2.6 represents the relationship between the heat of fusion and thermal conductivity for some of the most common materials used as PCMs for building applications. PCMs with low thermal conductivity (W/mk) are best suited for passive designing [43], such as: organic materials (Paraffin: RT27, Paraffin: RT18, Paraffin C18, all paraffins), inorganic materials (*n*-Heptadecane, *n*-Octadecane, all hydrated salts), and eutectics: (CA, CADE). Materials with higher thermal conductivity (W/Mk), from both solids and liquids, are capable for active designing [43], such as eutectics (CA-PA) and inorganic materials (Hydrated salt, $CaCl_2$ $6H_2O$, $CaCl_2$ $6H_2O$, L30, PCM Latest TM 25T TEAP Hydrated salt).

2.2.3 Benefits, Trends, Issues, and Opportunities

The main fields where PCMs are integrated in the building applications are floors and ceilings, walls, glazed systems, and air-based/free-cooling system. The benefits, trends, issues, and opportunities are shown in Table 2.6.

The major benefit of PCMs is to store a high energy density per unit of mass at a very near temperature range that can enhance indoor thermal comfort, improve thermal inertia of buildings, reduce temperature fluctuation, reduce building energy demand or reduce peak electricity demand for active cooling.

Table 2.6 Benefits, trends, issues, and opportunities of PCMs in building applications

	Benefits	Trends	Issues	Opportunities	References
Floors and ceiling	Small indoor temperature fluctuation; Enhancing indoor thermal comfort	Search for PCMs that can satisfy different climate regions and seasons	Does not work efficiently for all seasons, e.g., summer, winter	High energy storage densities; Avoid on-peak electric energy demand in case active heating/cooling system is used	[44]
Walls	Thermal inertia increase of lightweight structures; Enhancing thermal comfort; Reducing building energy demand for heating and cooling	PCM integration within bricks, concrete, gypsum boards, mortar	No full solidification; No full utilization of its latent heat; Low latent heat	Storing excess heat for later usage; Reducing building energy demand; Wall-temperature fluctuations reduction	[44, 45]
Glazed system	Energy saving of peak demand; Enhancing thermal comfort; Improved thermal inertia	Testing in various climate regions; Conceptual design testing for various seasons (summer, winter, etc.)	Low transparency; Container problem; Low latent heat; Large amount of energy dissipation through glazed system; Higher thermal conductivity of PCMs; Does not work efficiently for all seasons, e.g., summer, winter	Find a way for storing excess heat; Possibility to work as semiconductor for heat	[46–54]

(continued)

Table 2.6 (continued)

	Benefits	Trends	Issues	Opportunities	References
Air-based, free-cooling system	Applying PCMs in free-cooling systems increase the thermal storage	Effects of latent heat and melting temperature	Low latent heat; Low thermal conductivity of PCMs	High energy storage densities; Storing excess heat for later usage; Reducing energy demand and air-conditioning operational costs; Avoid on-peak electric energy demand	[44, 54, 55]

2.3 Remarks

In this research, the focus was on PCM thermal-physical properties suitable for a building application and benefits of integrated PCMs in building. Overall, the conclusion is that the PCM technology seems promising, but this is a relatively new area of research for applications in buildings. There are still some fields that need to be investigated for a large-scale application of this technology. Searching for new PCMs and technologies could be also important, for example, the possibilities of dynamically capable and controllable phase change temperature.

The major benefits of PCMs for buildings are high energy storage densities at near-constant temperature, low interior temperature variance, thermal comfort, and improved building thermal inertia of lightweight structures. Some issues associated are: partial solidification over the night, low thermal conductivity in case of PCM modulation, low transparency in case of glazing systems, container problem, still low latent heat of PCMs in the interval from 20 to 32 °C, and seasonal issue—inefficiency to operate interchangeably for heating and cooling seasons.

References

1. Rathod MK, Kanzaria HV (2011) A methodological concept for phase change material selection based on multiple criteria decision analysis with and without fuzzy environment. Mater Des 32(6):3578–3585
2. Telkes M, Raymond E (1949) Storing solar heat in chemicals—a report on the Dover house. Heat Ventilat 46(11):80–86
3. Sokolov M, Keizman Y (1991) Performance indicators for solar pipes with phase change storage. Sol Energy 47(5):339–346
4. Fath HE (1998) Technical assessment of solar thermal energy storage technologies. Renew Energy 14(1–4):35–40

5. Mettaweea E-BS, Assassab GM (2006) Experimental study of a compact PCM solar collector. Energy 31(14):2958–2968
6. Shuklaa A, Buddhib D, Sawhneya R (2009) Solar water heaters with phase change material thermal energy storage medium: a review. Renew Sustain Energy Rev 13(8):2119–2125
7. Kuznik F, David D, Johannes K, Jean-Jacques R (2011) A review on phase change materials integrated in building walls. Renew Sustain Energy Rev 15:379–391
8. Scalata S, Banua D, Hawesa D, Parishb J, Haghighataa F, Feldman D (1996) Full scale thermal testing of latent heat storage in wallboard. Solar Energy Mater Solar Cells 44(1):49–61
9. Shileia L, Guohuib F, Nenga Z, Li D (2007) Experimental study and evaluation of latent heat storage in phase change materials wallboards. Energy Build 39(10):1088–1091
10. Zhou G, Yang Y, Wang X, Zhou S (2009) Numerical analysis of effect of shape-stabilized phase change material plates in a building combined with night ventilation. Appl Energy 86(1):52–59
11. Lin K, Zhang Y, Xu X, Di H, Yang R, Qin P (2005) Experimental study of under-floor electric heating system with shape-stabilized PCM plates. Energy Build 37(3):215–220
12. Nagano K, Takeda S, Mochida T, Shimakura K, Nakamura T (2006) Study of a floor supply air conditioning system using granular phase change material to augment building mass thermal storage—heat response in small scale experiments. Energy Build 38(5):436–446
13. Lu T (2000) Thermal management of high power electronics with phase change cooling. Int J Heat Mass Transf 43(13):2245–2256
14. Kandasamy R, Wang X-Q, Mujumdar AS (2007) Application of phase change materials in thermal management of electronics. Appl Therm Eng 27(17–18):2822–2832
15. Tan F, Tso C (2004) Cooling of mobile electronic devices using phase change materials. Appl Therm Eng 24(2–3):159–169
16. Gin B, Farid MM (2010) The use of PCM panels to improve storage condition of frozen food. J Food Eng 100(2):372–376
17. Nomura T, Okinaka N, Akiyama T (2010) Waste heat transportation system, using phase change material (PCM) from steelworks to chemical plant. Resour Conserv Recycl 54(11):1000–1006
18. Ying B-A, Kwok Y-L, Li Y, Zhu Q-Y, Yeung C-Y (2004) Assessing the performance of textiles incorporating phase change materials. Polym Testing 23(5):541–549
19. Rathod MK, Banerjee J (2013) Thermal stability of phase change materials used in latent heat energy storage systems: a review. Renew Sustain Energy Rev 18:246–258
20. Zalba B, Marín JM, Cabeza LF, Mehling H (2003) Review on thermal energy storage with phase change: materials, heat transfer analysis and applications. Appl Therm Eng 23(3):251–283
21. Dinçer İ, Rosen MA (2002) Thermal energy storage—systems and applications. Wiley, Hoboken
22. Mehling H, Cabeza LF (2008) Heat and cold storage with PCM: an up to date introduction into basics and applications. Springer, Berlin
23. Sharma A, Tyagib VV, Chena CR, Buddhib D (2009) Review on thermal energy storage with phase change materials and applications. Renew Sustain Energy Rev 13(2):318–345
24. Ghoneim AA (1989) Comparison of theoretical models of phase-change and sensible heat storage for air and water-based solar heating systems. Sol Energy 42(3):209–220
25. Morisson A-K (1978) Effects of phase-change energy storage on the performance of air-based and liquid-based solar heating systems. Sol Energy 20(1):57–67
26. Sharma SD, Sagara K (2005) Latent heat storage materials and systems: a review. Int J Green Energy 2:1–56
27. Buddhi D, Sawhney R (1994) Proceeding of thermal energy storage and energy conversion. School of energy and environmental studies. In: Devi Ahilya University, Indore, India
28. Abhat A (1981) Development of a modular heat exchanger with an integrated latent heat storage. Report no. BMFT FBT 81-050, German Ministry of Science and Technology, Bonn
29. Sharma SD, Kitano H, Sagara K (2004) Phase change materials for low temperature solar thermal applications. Res Rep Fac Eng Mie Univ 29:31–64
30. Abhat A (1983) Low temperature latent heat thermal energy storage: heat storage materials. Sol Energy 30(4):313–332

31. Goia F, Boccaleri E (2016) Physical–chemical properties evolution and thermal properties reliability of a paraffin wax under solar radiation exposure in a real-scale PCM window system. Energy Build 119:41–50
32. Alkan C, Kaya K, Sari A (2009) Preparation, thermal properties and thermal reliability of form-stable paraffin/polypropylene composite for thermal energy storage. J Polym Environ 17(4):254–258
33. Hadjieva M, Kanev S, Argirov J (1992) Thermophysical properties some paraffins applicable thermal energy storage. Sol Energy Mater Sol Cells 27(2):181–187
34. Sharma S, Buddhi D, Sawhney RL (1999) Accelerated thermal cycle test of latent heat storage materials. Sol Energy 66(6):483–490
35. Shukla A, Buddhi D, Sawhney RL (2008) Thermal cycling test of few selected inorganic and organic phase change materials. RenewableEnergy 33(12):2606–2614
36. Lane GA (1983) Solar heat storage: latent heat materials, 1st edn. CRC Press, Florida
37. Hasnain SM (1998) Review on sustainable thermal energy storage technologies, part I: heat storage materials and techniques. Energy Conserv Manag 39(11):1127–1138
38. Hasan A, Sayigh AA (1994) Some fatty acids as phase-change thermal energy storage materials. Renew Energy 4(1):69–76
39. Sari A, Kaygusuz K (2003) Some fatty acids used for latent heat storage: thermal stability and corrosion of metals with respect to thermal cycling. Renew Energy 28(6):939–948
40. Alvia JZ, Imranb M, Peia G, Lia J, Gaoa G, Alvic J (2017) Thermodynamic comparison and dynamic simulation of direct and indirect solar organic Rankine cycle systems with PCM storage. 1(129):716–723
41. Estevesa LP, Magalhãesb A, Ferreirab V, Pinhoc C (2017) Evolution of global heat transfer coefficient on PCM energy storage cycles. Energy Procedia 1(136):188–195
42. Durakovic B, Yildiz G, Yahia ME (2020) Comparative performance evaluation of conventional and renewable thermal insulation materials used in building envelops. Tehnicki vjesnik—Technical Gazette 27(1) In Press
43. Miró L, Barreneche C, Ferrer G, Solé A, Martorell I, Cabeza LF (2016) Health hazard, cycling and thermal stability as key parameters when selecting a suitable. Phase Change Material (PCM) 1(16):1–20
44. Saffari M, Gracia Ad, Fernández C, Cabeza LF (2017) Simulation-based optimization of PCM melting temperature to improve the energy performance in buildings. Appl Energy 1(202):420–434
45. Edsjø S, Bjørn K, Jelle P (2015) Phase change materials for building applications: a state-of-the-art review and future research opportunities. Energy Build 1(1):1–16
46. Goia F (2012) Thermo-physical behaviour and energy performance assessment of PCM glazing system configurations: a numerical analysis. Front Archit Res 1(1):341–347
47. Durakovic B, Torlak M (2017) Simulation and experimental validation of phase change material and water. JMES 8(5):1837–1846
48. Durakovic B, Torlak M (2017) Experimental and numerical study of a PCM window model as a thermal energy storage unit. Int J Low-Carbon Technol 12(3):272–280
49. Liu C, Wu Y, Zhu Y, Li D, Ma L (2017) Experimental investigation of optical and thermal performance of a PCM-glazed unit for building applications. Energy Build 1(1):1–30
50. Liu Changyu, Wu Y, Li D, Zhou Y, Wang Z, Liu X (2017) Effect of PCM thickness and melting temperature on thermal performance of double glazing units. Journal of Building Engineering 11(1):87–95
51. Li D, Li Z, Zheng Y, Liu C, Hussein AK, Liu X (2016) Thermal performance of a PCM-filled double-glazing unit with different thermophysical parameters of PCM. Sol Energy 133(1):207–220
52. Li D, Ma T, Liu C, Zheng Yumeng, Wang Z, Liu X (2016) Thermal performance of a PCM-filled double glazing unit with different optical properties of phase change material. Energy Build 119(1):143–152
53. Duraković B, Mešetović S (2019) Thermal performances of glazed energy storage systems with various storage materials: an experimental study. Sustain Cit Soc 45:422–430

54. Torlak M, Delalić N, Duraković B, Gavranović H (2014) CFD-based assessment of thermal energy storage in phase-change materials—(PCM). In: International energy technologies conference proceedings—ENTECH'2014, Istanbul, Turkey
55. Xie J, Wang W, Liu J, Pan S (2018) Thermal performance analysis of PCM wallboards for building application based on numerical simulation. Solar Energy 1(162):533–540

Chapter 3
Passive Solar Heating/Cooling Strategies

3.1 Introduction

Global energy demand rapidly grows along with the population. Satisfying global energy demand will be challenging in the future. Application of passive building design techniques will play an important role in building energy demand reduction. Passive systems are uncomplicated and require minimal maintenance. Furthermore, it is a method of reducing the impact on the interior and exterior building environment and reduces costs. Passive design is practiced throughout the world and has been shown to be successful. Based on the case study from India, it was estimated that at least 35% of total building energy demand might be reduced using passive heating/cooling strategies [1].

3.1.1 Passive Design Factors

Passive strategies are influenced by various factors such as shape, size, and orientation. The orientation of a given building, especially its transparent surfaces, influences its energy consumption due to the different heights from which solar radiation hits. The proper orientation and dimensions of windows are dependent on outdoor climate and building purpose. Therefore, factors that affect energy consumption in buildings are: climate, urban context/geometry, building design, systems efficiency, and occupant behavior. Furthermore, the optimum arrangement of building structures may result in a saving between 10 and 20% for energy load used for air conditioning [2]. Performance of the building itself will show the impact on the design because layout and massing of the building can generate self-shading effects and enhance the ventilation and natural lighting. Building thermal performance can be achieved by any additional cost in adapting design strategies in a proper way. To minimize heat losses from building it requires minimization of the surface-to-volume ratio (*building shape coefficient*), which implies a reduction of the building envelope exposed

© Springer Nature Switzerland AG 2020

B. Duraković, *PCM-Based Building Envelope Systems*, Green Energy and Technology, https://doi.org/10.1007/978-3-030-38335-0_3

to the exterior environment. Reduced envelope exposure to the exterior environment reduces the availability of daylight and sunlight and increasing energy consumption for artificial lighting and natural ventilation. These two conflicting demands require looking for optimum solution for particular cases. *Building shape coefficient* may be a good indicator of building heating load. Studies explored the connection between shape coefficient and heating loads, which revealed that in the colder climates due to low solar gain in the glazing the heating load was proportional to the building shape coefficient. Therefore, to minimize heat loss from buildings, lowering value of building shape coefficient is desirable in early stage of the building design [3]. Another important indicator of energy consumption at conceptual stage of building design is passive volume ratio.

The following equations describe building shape coefficient and passive volume ratio. The building shape coefficient represents the ratio of total exterior building area to the built volume.

$$C_f = \frac{A_e}{V_b} \tag{3.1}$$

Minimizing C_f is desirable to achieve building energy demand reduction. Passive volume ratio represents the ratio between passive volume and built volume. Passive volume (*passive zone*) is defined as parts of a building which can be naturally lit and ventilated. Based on empirical data, all perimeter parts of buildings lying within 6 m of the façade (*or twice the ceiling height*) are considered as passive, while all the other zones are considered as non-passive. *Passive volume ratio* of buildings provides an estimate of the potential to implement passive and low energy techniques in conceptual design stage. However, in some cases passive zones of buildings may still be wastefully air-conditioned or artificially lit, which may consume more energy than non-passive zones. This may apply to excessive glazing ratios (*ratio between glazing area and gross exterior wall area*) and untreated façades, which makes them particularly vulnerable to overheating during the summer and to heat losses during the winter [4].

$$f_p = \frac{V_p}{V_b} \tag{3.2}$$

where C_f is building shape coefficient in m^{-1}, A_e is building envelope surface in m^2, V_b is built volume in m^3, V_p is passive volume in m^3, f_p is passive volume ratio ($0 < f_p < 1$).

Passive volume ratio is a good index at early stage of building design in assessing the passive design technologies and energy consumption. A higher passive design ratio is desirable, which provides an estimate of the potential to implement passive and low energy techniques. Ideally, passive design ratio should be equal to 1 ($f_p = 1$) [3].

In cold climates, implementation of the atrium due to bigger glazing to roof ratio in low-rise structures is preferable while in high-rise structures it is opposite, low

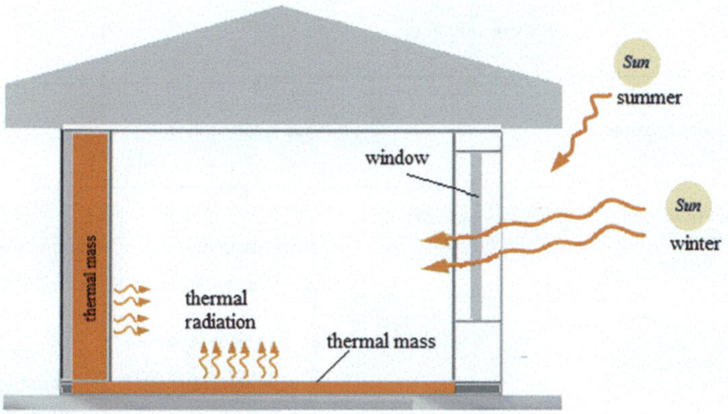

Fig. 3.1 Passive solar design strategies

glazing to roof ratio is preferable. The cube building shape is almost ideal shape for passive buildings, suitable for practical buildings and has the lowest building shape coefficient compared to other shapes. The shape provides good results for all climate types as well as attached units placed in the straight layout, while the detached units, L-shaped and U-shaped configuration, require extensive heating and cooling (the building shape coefficient is increased). The savings go up to 30% for cooling and 50% for heating, while 1–5% of energy saving can be achieved by changing the orientation and shape factors. Furthermore, increasing the south window size can significantly decrease the total annual load in colder climates, which will increase in warm conditions [2, 5] (Fig. 3.1).

Passive solar buildings refer to the use of solar energy for heating and cooling in buildings. The advantage of these concepts is cost reduction and impact on the environment. The system is not based on active heating/cooling systems, but usually has some movable parts that require minimal maintenance. Some concepts and results are discussed here. The result is mainly obtained from numerical studies and designing of experiments [6].

3.1.2 *Scope of the Chapter*

To select suitable passive heating/cooling technique, different factors discussed in the previous paragraphs have to be considered. This chapter discusses the classification and working principle of passive heating/cooling techniques as well as the summary of latest research results achieved. Classification of solar heating/cooling concept is shown in Fig. 3.2.

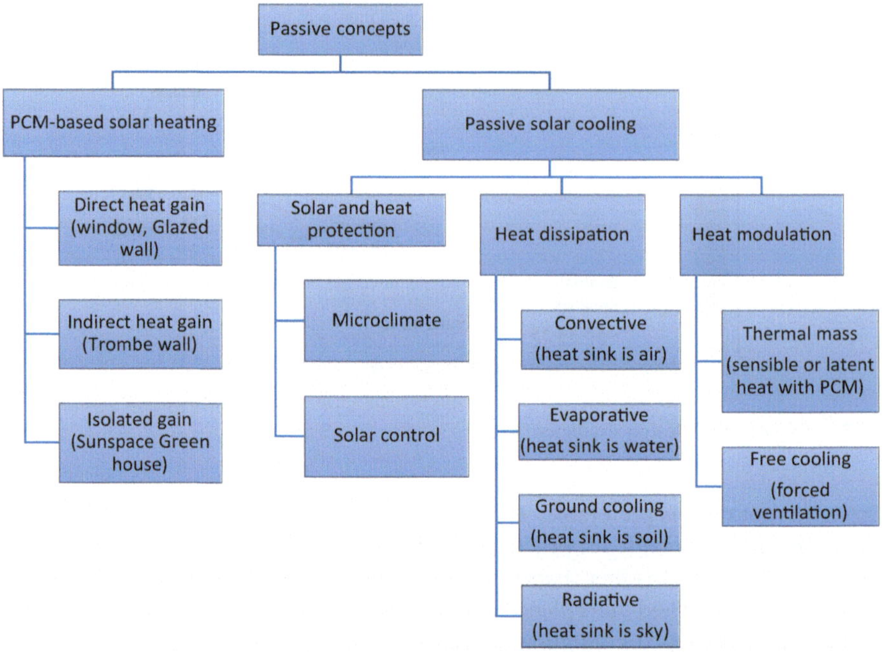

Fig. 3.2 Classification of passive solar heating/cooling concepts

Basically, the passive solar concepts can be classified as per its purposes such as passive concepts for solar heating, passive concepts for solar cooling, and combined passive concepts for solar heating and cooling. PCM-based solar heating concepts are based on thermal mass where the PCM acts as a heat sink and protects the building from overheating during the day, while over the night the heat is released from the PCM and heats the space.

Passive solar cooling concepts are categorized in three major groups such as heat dissipation, heat modulation, and heat protection. In heat dissipation technique, the excess heat is absorbed by environmental heat sinks such as air, water, soil, and sky. In heat protection, the building structure is protected from direct solar heat gain. In the heat modulation technique, the excess heat is absorbed by building thermal mass, which acts as the heat sink. The focus of this book is the PCM-based passive solar concepts.

3.2 Passive Solar Heating Concepts

Solar heating concept can be divided into direct solar heat gain, indirect solar heat gain, and isolated gain.

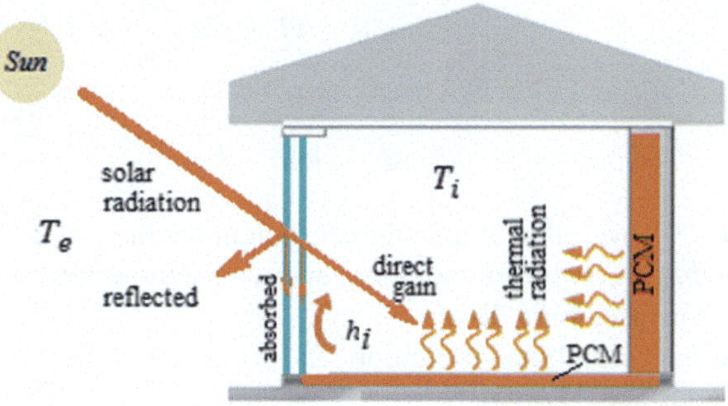

Fig. 3.3 Direct heat gain over daylight hours

3.2.1 *Direct Gain*

The simplest and the oldest traditional way of solar heating is direct heat gain through the south-facing glass pane. In direct solar heat gain during the day, building collects, absorbs, and then stores the heat, while during the night it releases the heat. Therefore, thermal heating is achieved by a glazed window in the living room which is directly transmitted with solar radiation as depicted in Fig. 3.3. This concept can be improved by applying the PCM to the floor. The PCM will enhance the energy storage capacity of the floor that will provide better heating over the night.

Studies have shown that using the south-oriented single-glazed system without using storage mass is inefficient, whereas, in contrast, using night insulation and soundly storage mass together with the double-glazed system is very efficient in meeting needed heating requirements. Solar heating storage capacity of 400 kJ/m^{2}gC for single- and double-glazed system and systems with night insulation shows that the percentage was about 10, 65, 80, and 90%, while being in 45° tilt that percentage was around 87, 80, 58, and 18% for the same systems, which means that the double-glazed system with night insulation is successful as an active system while being tilted near the optimum angle; therefore, it should be used to reduce heat loss from room, and during the night, windows should be covered by insulation to avoid similar problem. Since air is a poor conductor, the air gap helps reduce the heat transfer with conduction where reduction of heat gain is 9% and reduction of heat loss is 28% which is achieved by the double-glazed system [2].

3.2.2 *Indirect Gain*

The incident solar radiation on the glazing surface is divided into three basic components: reflected, transmitted, absorbed. Absorbed component is later emitted to the room and outside. The heat transfer rate for the direct heat gain can be expressed as sum of interior glass surface heat flux and transmitted radiation to the interior glass surface [7].

$$\dot{q} = tI + U\left(\frac{aI}{h_{\text{Re}}} + T_e - T_i\right) \tag{3.3}$$

$$\frac{1}{U} = \frac{1}{h_{\text{Ri}}} + \frac{1}{h_{\text{Re}}}; \quad U = \frac{h_{\text{Ri}} \cdot h_{\text{Re}}}{h_{\text{Ri}} + h_{\text{Re}}} \tag{3.4}$$

- U—overall heat transfer coefficient for the glass, W/m^{2}K
- h_{Re}—combined radiation and convection heat transfer coefficient at exterior surface, W/m^{2}K
- h_{Ri}—combined radiation and convection heat transfer coefficient at interior surface, W/m^{2}K
- I—solar intensity, W/m^2
- T_e, T_i—exterior and interior room temperature, °C
- t—glass transmittance
- a—glass absorptance
- r—glass reflectance.

With indirect solar heat gain, the heat is allowed to enter through glazing and it is stored within the thermal mass that is after transferred to the area through conduction and convection and thermal radiation. These indirect heat gain concepts are well known as Trombe wall (Fig. 3.4). For the traditional Trombe walls, the heat storage capacity of the wall depends on material thermal properties, surface area, and the volume of the material. Similarly, it is worth for the PCM-based Trombe wall. Since PCM-based Trombe wall uses latent heat storage, the storage capacity depends on the PCM latent heat, thermal conductivity of the area, and volume on the PCM.

The heat transfer rate to the interior for the indirect heat transfer is written as:

$$\dot{q} = tI + U\left(\frac{atI}{h_{\text{Re}}} + T_e - T_i\right) \tag{3.5}$$

French engineer Felix Trombe introduced the first solar heating device in 1956, while later patents were introduced on early 1970s by Anvar Trombe and early 1980s

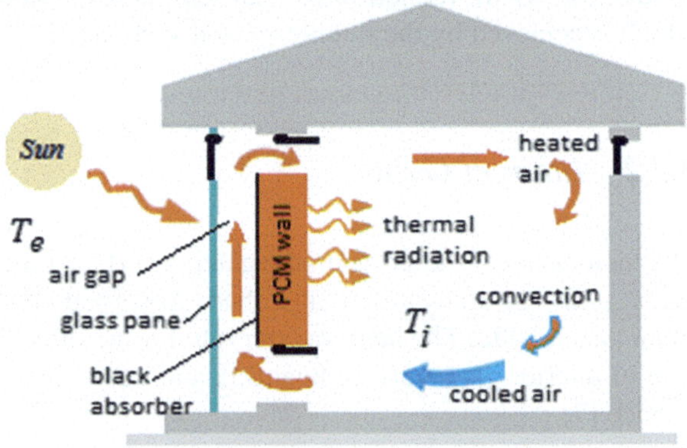

Fig. 3.4 Indirect heat gain

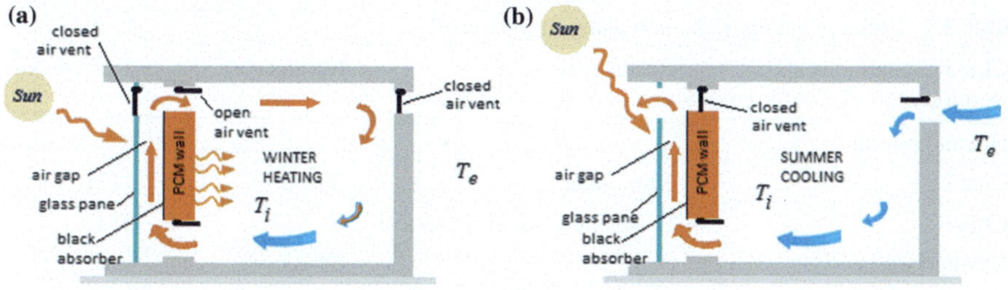

Fig. 3.5 Trombe wall for winter heating (**a**) and summer cooling (**b**)

[8]. The passive solar wall in various building elements was promoted in the 1960s [9]. The first house with solar wall was designed by architect Jacques Michel and built in Odeillo, France, in 1967 [10]. The Trombe wall has two vents for the air circulation. In winter case, both vents are opened allowing the air to enter at the bottom vent, passing through the gap between the wall and the glass pane, to enter the room through the upper vent, and to circulate in the room. By this way, stored heat in the wall is transferred to the interior by convection. Thermal radiation is another dominant mechanism of the heat transfer from the Trombe wall to the room [11]. In summer case, the upper vent above the Trombe wall facing the room is closed, while the bottom one is opened. Another vent on the glass pane at the top is opened allowing the hot air to leave the room due to stuck effect. On the opposite side of the room, there is another vent for the fresh air inlet (Fig. 3.5).

Passive solar buildings with Trombe walls are suitable for mixed and cold climates. Conventional heavy-weighted buildings with Trombe wall use thermal mass that is based on sensible heat storage. Sensible heat storage thermal mass increases volume and the weight of passive building. To reduce the weight and volume of the traditional passive buildings with Trombe wall, thermal mass is replaced with PCM. PCM-based Trombe wall stores the heat from exterior sources (solar heat) or interior sources such as active heating/cooling systems.

To foster greenhouse effect, Trombe wall is painted in black and covered with glazing thermal barriers. The heat is stored in PCM-based Trombe wall as latent heat and transmitted to the interior during the night by conduction through the wall, convection and thermal radiation.

There are several shortcomings of the Trombe wall such as uncertainty of the solar radiation caused unpredictable heating; heat transfer during the night from the interior to the exterior due to low resistance of the thermal mass; when the interior temperature is high, the air enters the room through the lower vent; reduces aesthetics due to painting the wall in black [12]. Trombe wall can be used for passive cooling as well. PCM-based Trombe wall absorbs the heat and acts as an insulator. Some results for building energy reductions using Trombe wall are shown in Table 3.1.

Table 3.1 Building energy reduction using Trombe wall

Climate conditions and location	Result type	Results (%)	References
Hot and humid			
China	Numerical	30–50	[13]
China	Experimental and numerical	51	[14]
Mediterranean		63–73	
Italy	Experimental and numerical	Single glazed 42	[15]
Italy	Experimental	Double glazed 48	[16]

3.2.3 *Isolated Gain*

In this concept, solar radiation is captured in a separate glazed solar space, which is usually next to the living space. The most practical way is to employ greenhouse, which can be attached later to the building for existing building or may be integrated into buildings for newly designed buildings. The heat can be stored simply within building, PCM-based structure, and can be distributed to the living space via ducts and vents. Sunspaces are suitable for passive heating in colder climates. This concept is actually a combination of direct and indirect heat gains in which the energy is stored in floors and walls, and by conduction through the wall and convection, it heats the living space. The solarium is constructed on south side of the building to catch as much as possible solar energy (Fig. 3.6).

This type of heating with solar gains is offered through windows in case the window is of high performance and is oriented toward the equator. According to studies, this reduces heating demand, uses natural daylight, and is environment friendly. The aim is to maximize solar gains during cooler periods. Window quilts are used to reduce thermal losses, as it was stated before, in the cooler periods. It consists of three sections—sunspace with the thick mass wall on the south side, linking space, and living space. Efficiency is achieved by the thermal wall among living space and sunspace which collects energy through the glazing and absorbs it to the living areas.

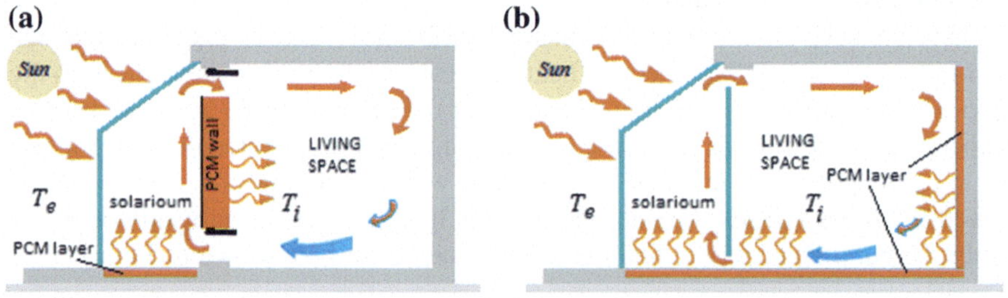

Fig. 3.6 Insulated gain—**a** combined PCM-based solarium with Trombe wall, **b** insulated and direct gain

3.3 Passive Solar Cooling Concepts

3.3.1 *Heat Dissipation*

3.3.1.1 Convective Cooling

In convective cooling, the heat sink is air. In this process, various modes of natural ventilation are used to remove the excess heat from the building. In the ventilation process, the buoyancy effect or natural wind speed is used as ventilation driving force. Buoyancy effect is used in solar wall and solar chimney, while wind-driven ventilation is used in cross ventilation.

The quality of indoor air is very important in the rooms where people spend a longer period of time. Poor quality of indoor air can temporarily result in discomfort, but also contribute to a serious long-term deterioration of occupant health. In standard houses, indoor air quality is maintained by ventilation of space [2]. Ventilation in building is required to ensure adequate air quality, to remove excess CO_2 and other harmful substances from the air, and to provide enough fresh air. In the room, every hour is required to provide 25–35 m^3 of fresh air per person. That means we should open the window or door for 15 min every three hours. The drainage of the extracted air from the room drains the heat, thus reduces the comfort of the room, and increases the need for heating. The passive house standard requires an exceptionally airtight building cladding, but in this way almost completely prevents fresh air from entering the room. For this reason, a ventilation device is installed in the passive house (Table 3.2).

Cross ventilation

The driving force in cross ventilation is pressure difference around the building. Positive pressure is created on the windward side of the building, while the negative pressure is created on the leeward side [21]. The pressure differences caused air from the high-pressure openings to low-pressure openings, achieving the cooling effect. Wind-induced ventilation uses pressures generated on the building by the wind, to

Table 3.2 Interior temperature reduction using convective cooling

Climate conditions and location	Result type	Results	References
Hot and dry			
UAE	Numerical and experimental	12 °C	[17]
Iran	Numerical	4 °C	[18]
Hot and humid			
India	Experimental	4 °C	[19]
China	Numerical	2 °C	[20]

Fig. 3.7 Cross ventilation

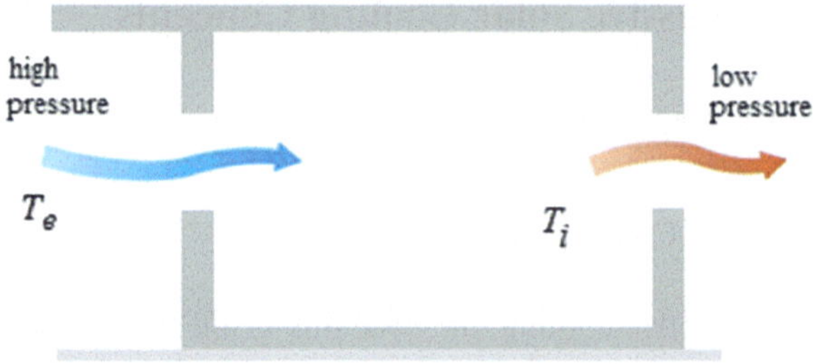

drive air through openings in the building, and leave on the opposite side, but can also drive single-sided ventilation and vertical ventilation flows. The working principle for opposite side is shown in Fig. 3.7.

Wind speed and direction are very variable. Openings must be controllable to cover the wide range of required ventilation rates and the wide range of wind speed. As with stack ventilation, the internal flow path inside the building must be considered. For cross ventilation, bear in mind that the leeward space will have air that has picked up heat or pollution from the windward space and thus may limit the depth of plan for cross ventilation. As with stack ventilation, the requirement for large openings may present problems with noise control. Also, the need to provide flow paths within the building may conflict with acoustic separation between internal spaces. However, the provision of by-pass ducts can help reduce this.

Solar chimney

To enhance natural ventilation in passive buildings, solar chimneys are good solutions. Solar chimney uses stack effect caused by temperature difference between chimney cavity and the interior of the building. Figure 3.8 shows typical solar chimneys implemented in passive building wall, roof, and combined solar chimney.

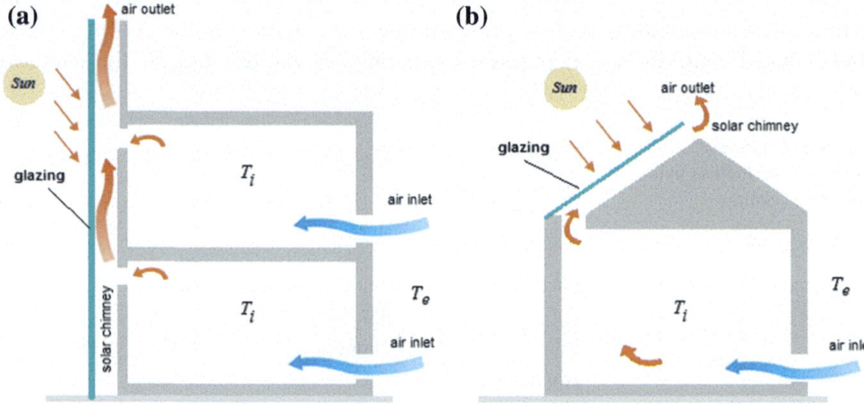

Fig. 3.8 Typical solar chimneys

Table 3.3 Building energy reduction using convective cooling

Climate conditions and location	Result type	Results	References
Hot and Humid		Energy reduction	
Thailand	Experimental	30%	[23]
Japan	Numerical	12%	[24]
Hot and dry		Temperature reduction	
Iran	Experimental	9–14 °C	[25]
Egypt	Experimental	8.5 °C	[26]

Within the heated room, there is the formation of a high-pressure area in the upper zone, with the tendency to leave the room in the direction of the area of lower pressure of the outer space. In the lower part forms a zone of reduced pressure—depression, with the tendency of entering the exhaled air of higher pressure. By contrast, in the days of the day when the indoor air temperature is often lower than the outside, there is an immediate opposite of the pressure distribution and air circulation process. To predict performances of solar chimneys, mathematical models have been utilized. These models are based on experimental or numerical results. More details on each mathematical model are provided in the literature such as ref number [22] (Table 3.3).

3.3.1.2 Evaporative Cooling

Evaporative cooling technology is a promising way to reduce building energy demand in providing fresh air to the occupant space. The working principle of this system is based on direct contact of hot and arid dry air with water. Hot and dry air streams over the water spray and the water evaporates by taking the heat from the air. As a result, the air becomes humid and cooler. In arid climates, the relative humidity of the air is raised to the range of 60–70%. In this process, the heat from the air is absorbed by sprayed water drops. The water drops are turned to vapor by increasing the humidity of the air and reducing the air temperature significantly. Therefore, evaporative cooling is suitable for hot and dry climates, but it is not suitable for humid climates since the air in these climates is almost saturated (Table 3.4).

Infiltration/ventilation

There are two types of airflows within the building which are caused by the pressure differences that lead to the natural airflow. The first one is the random openings such as the interfaces, crack, and gaps within the building. These openings cause infiltration, and to prevent them, we could use draft sealing, airlocks, airtight, and high-quality doors and windows. According to Wang et al., heating/cooling performance is effected by airtightness. Because of the hot and humid air, the cooling load

Table 3.4 Building energy reduction using evaporative cooling

Climate conditions and location	Result type	Results	References
Hot and dry		Energy reduction	
Iran	Experimental	>60%	[27–29]
India	Analytical	55%	[30]
Oceanic			
England	Experimental	6–8 °C	[31]
Hot and dry		Temperature reduction	
India	Experimental	5–8 °C	[31, 32]
China	Experimental	9–14 °C	[33]

and the latent load were increased by 9.4–56%. Ventilation using wind towers is the other airflow type. To provide the airflow within the building, windows are very important for the air circulation. The optimum air movement is 0.2–0.4 m/s for every season. Combining evaporative cooling with ventilation, it is possible to reduce temperature up to 17 °C, which is one of the most economical ways of building cooling [2].

At the top of the tower, cooler pads are placed. Water passes down through the pads and collects in the water tank. Air goes through wet pads at the top of the tower, and water evaporates and cools the air. The air circulation is based on wind and buoyancy. The air circulation can be enhanced with solar chimney on the other side. The sun heats the chimney causing heated air to go up, while on the other side through the wind tower there is an intake of the air from the exterior, cooling the air in the tower and entering into the living space (Fig. 3.9).

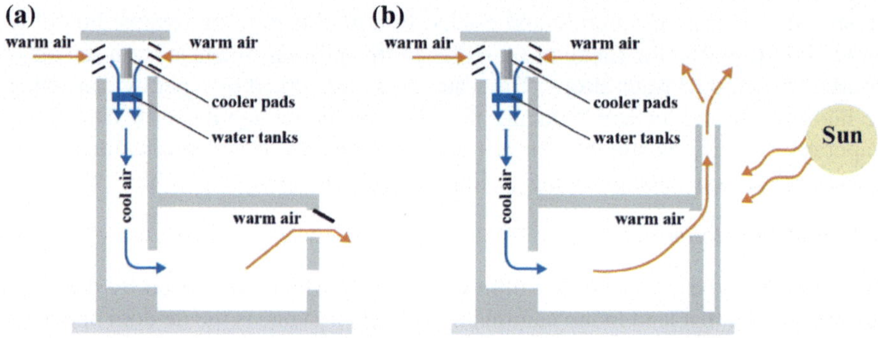

Fig. 3.9 Wind tower, **a** basic, **b** with solar chimney

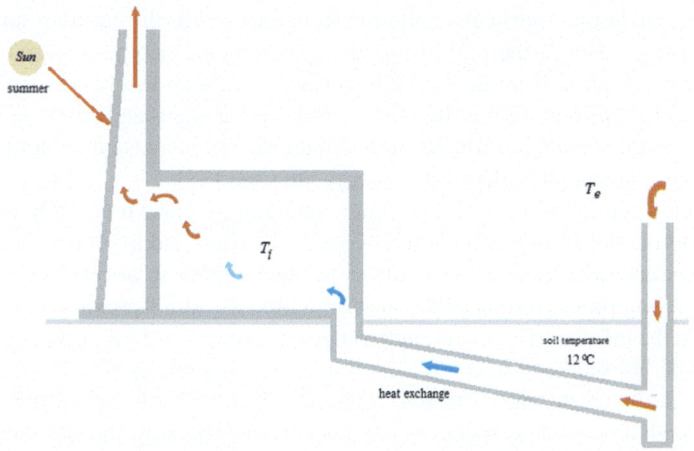

Fig. 3.10 Earth to air heat exchange

3.3.1.3 Ground Cooling

Earth to air heat exchange

In ground cooling, the soil is used as the heat sink. The earth soil is warmer in the winter when compared with the outside (above the soil) temperature. But in the summer the earth is cooler when compared to the outside temperature (above the soil). The soil has average temperature of about 12 °C over whole year. Therefore, the soil has the potential to cool in summer and to preheat the air in winter (Fig. 3.10).

This is the most common technique to use the soil as heat sink for cooling of buildings. The system uses pipes that are buried to depth of 1–3 m underground. The air circulates through the pipes and enters into the interior as cooled [2]. The best example is ancient Persian architecture. Ground cooling works well in combination with solar chimney. The air in the chimney is heated and moved upward by buoyancy forces, at the same time sucking fresh air from the exterior through the underground pipes and supplying to the interior.

3.3.1.4 Radiative Cooling

In radiative cooling, the heat sink is sky. Radiative cooling concepts are based on the application of materials that have high solar reflectance and high infrared emittance. Combining these two material properties, it is possible to reduce solar heat gain to building and enhance heat removal from the building by longwave radiation. It is one of the developing areas of passive cooling systems with a significant cooling potential. The method depends on the mechanism of radiative heat exchange [34]. The system depends on the difference between the shortwave radiations coming from sun and the longwave radiation coming from objects on earth, such as the buildings.

During daylight hours, shortwave radiation from the sun hits the building surface and heats the surface. But at the night-time when there is no incoming solar radiation, the longwave radiation from the building surface toward atmosphere is sufficient to achieve cooling. Because of this difference between the night- and daytime, radiative cooling has been considered for the night-time only. To increase the cooling effect, building materials which reflect solar energy are used [34].

However, scientific developments in nanotechnology made it possible to achieve daytime cooling below the ambient temperature. Designs and microstructures based on nanophotonic principles not only allow to reach higher solar reflection, but also generate cooling power through IR emission within the atmospheric window [35].

Between the infrared (IR) wavelength range between 8 and 13 μm, the atmosphere has a transparent window, which means it has very low absorption of the radiation coming from the earth, thus very low reflection back to earth. When the radiation passes through this window, it goes to the outer space. This way, the ambient temperature of the earth can be cooled via radiative emission through atmospheric window to the outer space behind [35].

Based on this, materials specifically designed to absorb and emit radiation from earth surface to the sky between 8 and 13 μm create a negative heat difference between the coming in solar radiation and going out surface radiation. Nanotechnology today opens the way for creating such materials aiming at emitting heat from the surface in its extremity within the atmospheric window, transporting higher amount of heat from the earth to the outer space, thus creating cooling day or night [34]. In addition to the solar radiation coming from the sun, there are other factors that affect the performance of this cooling system, such as the nonradiative (conductive and convective) heat created by objects on the earth and the emission type of the radiator (broadband or selective) used [35].

Mathematically speaking, the heat exchange mechanism can be calculated as the net cooling power.

$$P_{\text{net}} = P_R - P_a - P_{\text{nonrad}} - P_{\text{solar}} \tag{3.6}$$

where P_R is the emitted radiative power by the radiator; P_a is the atmospheric radiation absorbed by the radiator; P_{nonrad} nonradiative heat coefficient; P_{solar} is the absorbed solar power by the radiator [35]. Figure 3.11 shows the basic principle of a radiative cooling system.

Heat exchange at night between the radiator and the sky can be expressed mathematically as follows:

$$\dot{q}_R = \varepsilon\sigma\left(T_{\text{sky}}^4 - T_{\text{radiator}}^4\right) \tag{3.7}$$

where ε is emissivity, σ is Stefan–Boltzmann constant which is 5.670367×10^{-8} kg/s^{3}K^4, T_{sky}, temperature of the sky in K, T_{radiator}, temperature of the radiator surface in K (Table 3.5).

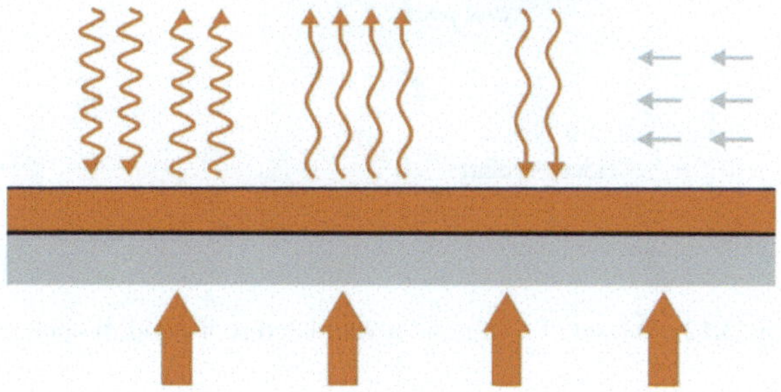

Fig. 3.11 Working principle of radiative cooling system

Table 3.5 Building temperature reduction using roof radiative cooling

Climate conditions and location	Result type	Results	References
Hot and humid			
Norway	Experimental	1–4 °C	[36]
Thailand	Experimental	1–6 °C	37[]
Canada	Experimental	6–20 °C	[38]
Warm and temperate			
Greece	Experimental	2.5–4 °C	[39]

3.3.2 *Solar and Heat Protection*

Roof pond

The heat gain of a single-story building is 50%, which is gained from the roof. To reduce this heat flux, numerous ways are used such as false ceilings, insulations [40], increasing the roof thickness, roof shading, and roof coatings. The roof temperature rises up to 65 °C without any special treatment. Buildings that have installed roof pond with the depth of the pond in a range of 0.05–0.15 m, can reduce the roof temperature to 40 °C [41]. The roof ponds are recommended for arid climates. The working principle of a roof pond is shown in Fig. 3.12.

Working principle is based on vapor pressure difference between the vapor at the water surface and the vapor in the surrounding air. The advantage of the roof pond is to achieve highest cooling potential with least maintenance (Table 3.6).

Courtyard planning

To reduce building energy demand, microclimate and thermal interaction of the exterior and interior of building may play an important role. Trees and green vegetation around the building are cheap solution to reducing building solar heat gain.

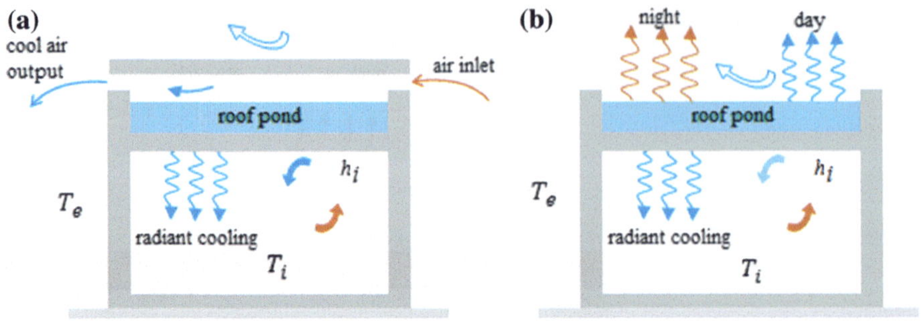

Fig. 3.12 Variant of roof pond **a** ventilated roof pond, **b** open roof pond

Table 3.6 Building energy demand reduction based on experiments and numerical simulation

Climate conditions and location	Result type	Results	References
Hot and dry		Energy reduction	
India	Numerical	35%	[42]
Iran	Numerical	79%	[43]
Warm and temperate			
South Africa	Numerical	59%	[44]
Hot and dry		Temperature reduction	
Israel	Experimental	4 °C	[45]
Iraq	Experimental	6.5 °C	[46]

Working principle is based on building shading and absorbing the heat by microclimate through the evapotranspiration process. In evapotranspiration process, the water from the green vegetation around the building is transferred to the atmosphere by evaporation. In this process of evaporation, the heat from the air in microclimate is absorbed as latent heat, contributing to reducing microclimate air temperature. In the past, water fountains were put and greeneries were planted at the courtyards, which enhanced the humidity and evaporative cooling, and shading [2] (Fig. 3.13).

Including green vegetation in building interiors such as atria, indoor plants or green roofs, green terrace, heat prone surfaces can have a significant impact on reducing building energy demand [47] (Table 3.7).

Solar shading techniques

Solar shading techniques may reduce the solar heat gain, indoor temperature, and the cost of cooling. The technique may provide interior temperature reduction of 2.5–4.5 °C. If insulation is used together with the above-mentioned technique, the indoor temperature can be reduced to 4.4–6.8 °C [53].

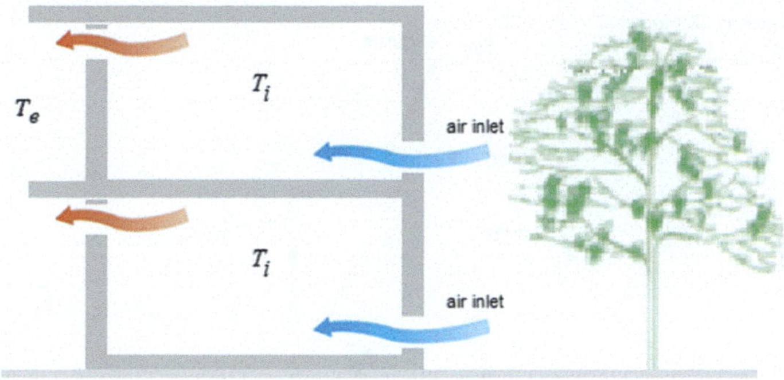

Fig. 3.13 Courtyard planning

Table 3.7 Building energy demand reduction due to vegetation effect

Climate conditions and location	Result type	Result (%)	References
Hot and dry			
USA	Numerical	20	[48]
Hot and humid			
USA	Numerical	6.1–14.4	[49, 50]
Malaysia	Experimental	29	[51]
China	Numerical	10.3	[52]

Shading devices are installed on the exterior, interior, and in some conditions both interior and exterior. The purpose is to reduce solar heat gain in summer and increase solar heat gain in winter. Consequently, the cooling/heating load might be reduced, which has positive effect on operating costs. Solar shading devices are divided into two groups: interior shading devices and exterior shading devices. The interior shading devices limit the solar radiation that comes from the sun; they have adjustable angles to allow or eliminate light to enter the interior. The interior shading can be either vertical (e.g., *vertical slats*) or horizontal (e.g., *Venetian blind*) that is installed on windows. Exterior shading devices can be vertical or horizontal and have some advantages over interior shading devices such as functionality, no late heat gain to the interior in summer. The horizontal shading devices prevent the excess light to enter the interior in summer and enable the light to enter the room in the winter which may heat the room into a warm place. The vertical shading devices are mounted on the east and west orientation; they improve insulation and act like a windbreak.

An important element for passive solar design is overhang limits that may control the solar energy to the place the user wants (Fig. 3.14).

The sun path is different during different seasons. During the summer, the sunlight directly enters into the room, but in winter the sunlight enters less into the room. Therefore, in the summer, the building is fully shaded which keeps the building's

Fig. 3.14 Shading devices with an overhang

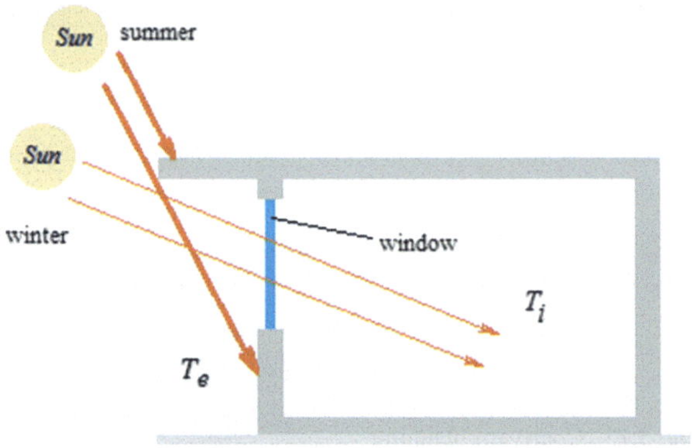

thermal mass in shade that cools the building, and in the winter, the building receives the full sunlight. The material density effects the temperature variation, and it is chosen accordingly to keep the room in stable temperature (Table 3.8).

Windows

Windows, shape, orientation, and scale of the windows are important in terms of energy reduction process. Direction of the wind has to be taken into account as well window to wall ratio (i.e., if it is close to the floor, its efficiency is poor; if it is closer to the middle of the wall, it is more efficient). Since air is circulating in the building, air leaks are to be taken into account due to heat loss possibility. To reduce convection, low e-glass windows with double glaze can be installed. Heat gain reduction is permissible with PCM integrated with glazing system [58, 59].

Convection is when the heat energy is transferred by moving fluids. Air is always circulating in the building. Therefore, air leaks cause up to 40 percent of the building heat loss. To reduce the convection, argon-filled (low e-glass) windows with double glazing are a good solution. Triple glazed window could be used as well but the problem might be the weight of glass [60] (Table 3.9).

Light shelves

Light shelves prevent excessive light and heat at the same time. The sunlight that bounces in the room from the shelf mounted on the ceiling of the room to the window

Table 3.8 Building energy demand reduction with shading techniques

Climate conditions and location	Result type	Result (%)	References
Hot and humid			
Taiwan	Numerical	11.3	[54]
Belgium	Numerical	12	[55]
Singapore	Experimental	2.62–10.13	[56]
Taiwan	Experimental	25	[57]

Table 3.9 Building energy demand reduction with various glazing techniques based on numerical results

Climate conditions and location	Glazing type	Result (%)	References
Hot and humid			
Malaysia	Triple glazed window	5.5–8.5	[61–63]
USA	Electrochromic glazing	20	[64]
Warm			
Italy	Low e-double glazing	52	[65]
Hot and cold			
China	Thermotropic glazing	2.4–12.3	[66–68]

then bounces back into the room. If the room height is large, it works better. If a corrugated reflective surface is used, the reflection at high altitude can be maximized. The premises operate effectively in the north and in the south (Fig. 3.15).

Interior lighting shelves split the windows between the visible part and the part that allows additional natural light to bounce up and reflect sunlight from the ceiling so that it can penetrate deeper into the floor panels. Exterior sun shades also direct daylight to the interior ceiling of the building, so they are classified as a type of shelf, but their primary purpose is to provide exterior shade for windows that generally reduce glare, increase homogeneity, and reduce solar heat gain. The ratio of the windows to the wall, the size and position of the windows, as well as building orientation are very important factors in building energy demand reduction [3].

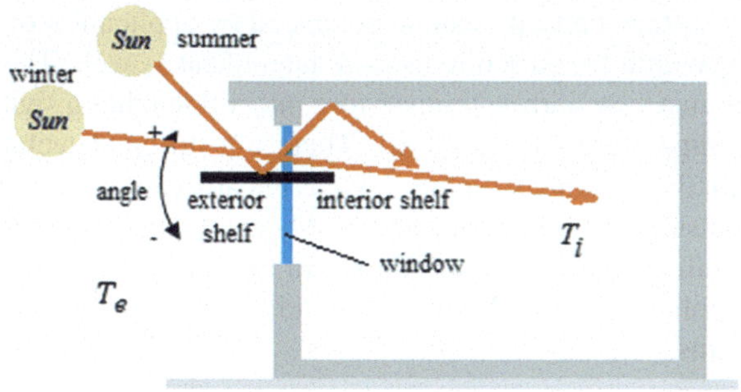

Fig. 3.15 Working principle of light shelves

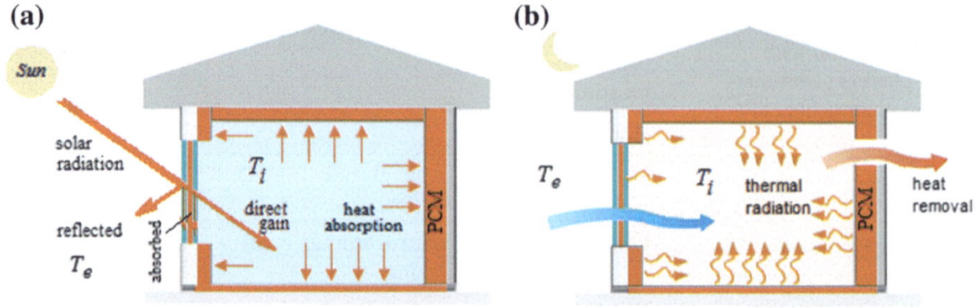

Fig. 3.16 Thermal mass working principle: **a** daytime heat absorption, **b** at night heat removal

3.3.3 Heat Modulation

In heat modulation technique, the heat sink is the thermal mass of the building structure. The heat gain is reduced by application of material with enhanced thermal storage capacity within building structure. Thermal mass and free cooling are two basic principles of heat modulation.

Thermal mass

Thermal mass is an effective way to reduce interior air temperature variation by storing the excess heat within the building structure and components and increasing building thermal inertia [58]. There are many factors that have impact on the efficiency of the building with enhanced thermal mass such as climate condition and building orientation, and construction material properties [69]. The heat transfers quickly to the interior and releases back to the exterior in buildings with low thermal mass, which is not desirable (Fig. 3.16).

Building construction materials have limited storage capacity due to its sensible heat storage. The storage capacity can be enhanced by application of phase change materials (PCMs) within the structure to store latent heat as well. The heat is stored and released back in a very narrow temperature interval that has a significant impact on the interior temperature variance [70]. There are different techniques to integrate PCMs within the building structure such as encapsulation, immersion, vacuum impregnation, and shape stabilization. The PCMs can be integrated into wallboard, brick, concrete, ceilings and roofs, and floors. In the following chapter will be presented in detail application of PCMs within building structure and components and free cooling systems.

Free cooling with active system

Free cooling is a technique of reducing interior building air temperature using naturally cool air instead of mechanical air-conditioning system (Fig. 3.17). The system is not cost-free since it uses fans or pumps for fluid circulation and required periodic maintenance. During the day, the heat gain of the building is stored and released outside by circulation of colder exterior air through the building. In this ventilation

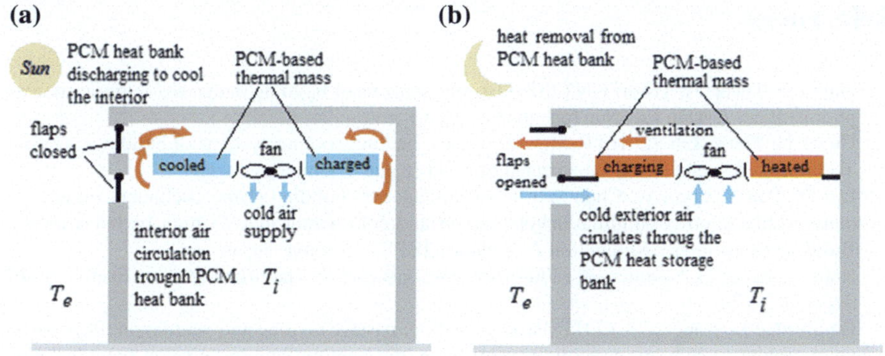

Fig. 3.17 Working principle of free cooling system: **a** over daytime PCM heat bank discharging, **b** overnight PCM heat bank charging

process, the building structure is cooled that helps reduce the heat gain during the next day. In this process, cold energy during the night is accumulated in thermal mass (a cold storage bank) to be utilized the next day [71]. Thermal mass of the storage bank is limited by its size and material properties, while the cooling potential is limited by the exterior air temperature [71].

3.4 Remarks

Passive techniques have potential to reduce building energy demand and help keeping interior temperature within comfort zone.

PCM-based heating with Trombe wall is convenient for heating and cooling in hot and cold climates in terms of energy performances and thermal comfort.

Heat control technique is efficient for avoiding direct heat gain to the interior in hot climate areas where there is plenty of space available for construction. Factors like aesthetics and structural aspect of a building play an important role due to extra mass added to the building.

Heat dissipation is suitable in the areas where the heat can be removed with presented fluid. The heat sink in this process can be air, soil, water, and sky in case of radiative cooling. This technique is suitable in cases where the building thermal mass is poor or in cases where it is not possible to improve it. The technique can be enhanced using suitable heat exchangers.

The heat modulation technique is based on the PCMs, and it is more suitable for the hot climate areas where the solar heat gain is intense. The most important factors for this technique are stability of PCMs and sustainability of the containers to prevent leakage of PCMs. In this technique, PCMs can be integrated into building structure, components, or in separate modules, which is discussed in detail in the following chapters.

References

1. Gupta N, Tiwari A, Tiwari GN (2016) Exergy analysis of building integrated semitransparent photovoltaic. Jestech Eng Sci Technol Inte J 20:3
2. Gupta N, Tiwari GN (2016) Review of passive heating/cooling systems of buildings. Global Technology Watch Group, Department of Science and Technology, India
3. Lin M, Pan Y, Long W, Chen W (2014) Influence of building shape coefficient on energy consumption of office buildings in hot-summer-and-cold-winter area of China. In: International Building Performance Simulation Association IBPSA, Nagoya, Japan
4. Ratti C, Baker N, Steemers K (2005) Energy consumption and urban texture. Energy Build 37(7):762–776
5. Bainbridge D, Haggard K (2011) Passive heating, passive cooling and ventilation. In: Passive solar architecture, heating, cooling, ventilation, daylighting, and more using natural flows. White River Junction, Vermont, Chelsea Green Publishing Company, Chaps. 2 and 3
6. Durakovic B (2017) Design of experiments application, concepts, examples: state of the art. Period Eng Nat Sci 5(3):421–439
7. Yellott JI (1963) Calculation of solar heat gain through single glass. Sol Energy 7(4):167–175
8. Morse E (1881) Warming and ventilating apartments by the sun s rays. U.S. Patent 246626 A, 11 Apr 1881
9. Omrany H, Ghaffarianhoseini A, Ghaffarianhoseini A, Raahemifar K, Tookey J (2016) Application of passive wall systems for improving the energy efficiency in buildings: a comprehensive review. Renew Sustain Energy Rev 62:1252–1269
10. Medici P (2018) The Trombe Wall during the 1970s: technological device or architectural space? Critical inquiry on the Trombe Wall in Europe and the role of architectural magazines. SPOOL 4(2):45–60
11. Rabani M, Kalantar V, Rabani M (2017) Heat transfer analysis of a Trombe wall with a projecting channel design. Energy 134:943–950
12. Saadatian O, Sopian K, Lim C, Asim N, Sulaiman M (2012) Trombe walls: a review of opportunities and challenges in research and development. Renew Sustain Energy Rev 16:6340–6351
13. Yang H, Burnett J, Ji J (2000) Simple approach to cooling load component calculation through PV walls. Energy Build 31(3):285–290
14. Sun W, Ji J, Luo C, He W (2011) Performance of PV-Trombe wall in winter correlated with south façade design. Appl Energy 88(1):224–231
15. Mastrucci A, di Perna C, Stazi F (2012) Trombe wall management in summer conditions: an experimental study. Solar Energy 86(9):2839–2851
16. Stazi F, Mastrucci A, Di C (2012) Perna the behaviour of solar walls in residential buildings with different insulation levels: an experimental and numerical study. Energy Build 47(April):217–229
17. Calautit J, Hughes B (2016) A passive cooling wind catcher with heat pipe technology: CFD, wind tunnel and field-test analysis. Appl Energy 162(1):460–471
18. Ghadiri M, Ibrahim N, Dehnavi M (2011) The effect of tower height in square plan wind catcher on its thermal behavior. Aust J Basic Appl Sci 5(9):381–385
19. Priya RS, Sundarraja M, Radhakrishnan S (2012) Experimental study on the thermal performance of a traditional house with one-sided wind catcher during summer and winter. Energ Effi 5(4):483–496
20. Ji Z, Su Y, Khan N (2012) Performance evaluation and energy saving potential of windcatcher natural ventilation systems in China. Int J Archit Eng Constr 1(2):84–95
21. ASHRAE Handbook of Fundamentals (2009) USA: American society of heating, refrigerating and air-conditioning engineers, Inc.
22. Shi L, Zhang G, Yang W, Huang D, Cheng X, Setunge S (2018) Determining the influencing factors on the performance of solar chimney in buildings. Renew Sustain Energy Rev 88(May):223–238

23. Khedari J, Rachapradit N, Hirunlabh J (2003) Field study performance of solar chimney with air-conditioned building. Energy 28(11):1099–1114
24. Miyazaki T, Akisawa A, Kashiwagi T (2006) The effects of solar chimneys on thermal load mitigation of office buildings under the Japanese climate Renew. Renew Energy 31(7):987–1010
25. Rabani R, Faghih A, Rabani M, Rabani M (2014) Numerical simulation of an innovated building cooling system with combination of solar chimney and water spraying system. Heat Mass Transf 50(11):1609–1625
26. Amer E (2006) Passive options for solar cooling of the buildings in arid areas. Energy 31(8–9):1332–1344
27. Heidarinejad G, Bozorgmehr M, Delfani S, Esmaeelian J (2009) Experimental investigation of two-stage indirect/direct evaporative cooling system in various climatic conditions. Build Environ 44(10):2073–2079
28. Delfani S, Esmaeelian J, Pasdarshahri H, Karami M (2010) Energy saving potential of an indirect evaporative cooler as a pre-cooling unit for mechanical cooling systems in Iran. Energy Build 42(11):2169–2176
29. Hajidavalloo E (2007) Application of evaporative cooling on the condenser of window-air-conditioner. Appl Therm Eng 27(11–12):1937–1943
30. Jain J, Hindoliya D (2016) Energy saving potential of indirect evaporative cooler under Indian climates. Int J Low-Carbon Technol 11(May):193–198
31. Ibrahim E, Shao L, Riffat S (2003) Performance of porous ceramic evaporators for building cooling application. Energy Build 35(9):941–949
32. Bishoyi D, Sudhakar K (2017) Experimental performance of a direct evaporative cooler in composite climate of India. Energy Build 153(10):190–200
33. Wu J, Huang X, Zhang H (2009) Numerical investigation on the heat and mass transfer in a direct evaporative cooler. Appl Therm Eng 29(1):195–201
34. Burnett M (2015) Passive Radiative Cooling. Stanford University, California
35. Gu M, Hossain M (2016) Radiative cooling: principles, progress, and potentials. Adv Sci 3(7):1500360
36. Meir M, Rekstad J, LØvvik O (2002) A study of a polymer-based radiative cooling system. Solar Energy 73(6):403–417
37. Khedari J, Waewsak J, Thepa S, Hirunlabh J (2000) Field investigation of night radiation cooling under tropical climate. Renew Energy 20(2):183–193
38. Hollick CJ (2012) Nocturnal radiation cooling tests. Energy Proc 30:930–936
39. Bagiorgas H, Mihalakakou G (2008) Experimental and theoretical investigation of a nocturnal radiator for space cooling. Renew Energy 33(6):1220–1227
40. Durakovic B, Yildiz G, Yahia ME (2020) Comparative performance evaluation of conventional and renewable thermal insulation materials used in building envelops. *Tehnicki vjesnik—Technical Gazette* 27(1) (in press)
41. Ayyoob Sharifi YY (2015) Roof ponds as passive heating and cooling systems: a systematic review. Appl Energy 160:337–357
42. Sodha M, Kumar A, Singh U, Tiwari GN (1980) Periodic theory of an open roof pond. Appl Energy 7(4):305–319
43. Raeissi S, Taheri M (1996) Cooling load reduction of the buildings using passive roof options. Renew Energy 7(3):301–313
44. Vorster J, Dobson R (2011) Sustainable cooling alternatives for buildings. J Energy South Afr 22(4):48–66
45. Runsheng T, Etzion Y, Erell E (2003) Experimental studies on a novel roof pond configuration for the cooling of the buildings. Renew Energy 28(10):1513–1522
46. Kharrufa S, Adil Y (2008) Roof pond cooling of the buildings in hot arid climates. Build Environ 43(1):82–89
47. Raji B, Tenpierik MJ, Dobbelsteen A (2015) The impact of greening systems on building energy performance: a literature review. Renew Sustain Energy Rev 45(5):610–623

48. Akbari H, Pomerantz M, Taha H (2001) Cool surfaces and shade trees to reduce energy use and improve air quality in urban areas. Sol Energy 70(3):295–310
49. Simpson J (2002) Improved estimates of tree-shade effects on residential energy use. Energy Build 34:1067
50. Pandit R, Laband D (2010) Energy savings from tree shade. Ecol Econ 69(6):1324–1329
51. Fairuz M, Jones P, Gwilliam J, Salleh E (2012) An evaluation of outdoor and building environment cooling achieved through combination modification of trees with ground materials. Build Environ 58:245–257
52. Hsieh C, Li J, Zhang L, Schwegler B (2018) Effects of tree shading and transpiration on building cooling energy use. Energy and Buildings 159(1):382–397
53. Bhamare DK, Rathod MK, Banerjee J (2019) Passive cooling techniques for building and their applicability in different climatic zones—the state of art. Energy Build 198(1):467–490
54. Yu J, Yang C, Tian L (2008) Low-energy envelope design of residential building in hot summer and cold winter zone in China. Energy Build 40(8):1536–1546
55. Tzempelikos A, Athienitis A (2007) The impact of shading design and control on building cooling and lighting demand. Sol Energy 81(3):369–382
56. Wong N, Li S (2007) A study of the effectiveness of passive climate control in naturally ventilated residential buildings in Singapore. Build Environ 42(3):1395–1405
57. Yang K, Hwang R (1995) Energy conservation of the buildings in Taiwan. Pattern Recogn 28(10):1483–1491
58. Durakovic B, Torlak M (2017) Experimental and numerical study of a PCM window model as a thermal energy storage unit. Int J Low-Carbon Technol 12(3):272–280
59. Durakovic B, Torlak M (2017) Simulation and experimental validation of phase change material and water used as heat storage medium in window applications. J Mater Environ Sci 8(5):1746–1837
60. Smith P (2005) Lighting—designing for daylight. In Architecture in a climate of change. Architectural Press, pp 181–187
61. Tahmasebi M, Banihashemi S, Hassanabadi M (2011) Assessment of the variation impacts of window on energy consumption and carbon footprint. Proc Eng 21:820–828
62. Sadrzadehrafiei S, Sopian K, Mat S, Lim C, Hashim H, Zaharim A (2012) Potential energy and emission reduction through application of triple glazing. In: Proceedings of the 6th international conference on energy and development—environment—biomedicine (EDEB12), pp 138–142
63. Sadrzadehrafiei S, Sopian K, Mat S, Lim C (2011) Application of advanced glazing to mid-rise office buildings in Malaysia. In: Proceedings of WSEAS conference EED
64. Sbar N, Podbelski L, Yang H, Pease B (2012) Electrochromic dynamic windows for office buildings. Int J Sustain Built Environ 1(1):125–139
65. Buratti C, Moretti E (2012) Experimental performance evaluation of aerogel glazing systems. Appl Energy 97(9):430–437
66. Yao J, Zhu N (2012) Evaluation of indoor thermal environmental, energy and daylighting performance of thermotropic windows. Build Environ 49(3):283–290
67. Li P, Guo W, Qiu Z (2011) Energy performance of PV windows used in different cities of China. In: ASME 2011 power conference collocated with JSME ICOPE2011, American Society of Mechanical Engineers
68. Guo W, Qiu Z, Li P, He J, Zhang Y, Li Q (2012) Application of PV window for office building in hot summer and cold winter zone of China. In: 2011 international conference on energy, environment and sustainable development, ICEESD 2011, Shanghai
69. Duraković B, Mešetović S (2019) Thermal performances of glazed energy storage systems with various storage materials: an experimental study. Sustain Cities Soc 45(2):422–430
70. Torlak M, Delalić N, Duraković B, Gavranović H (2014) CFD-based assessment of thermal energy storage in phase-change materials—(PCM). In: International energy technologies conference proceedings—ENTECH'2014, Istanbul, Turkey
71. Zeinelabdein R, Omera S, Gan G (2018) Critical review of latent heat storage systems for free cooling in buildings. Renew Sustain Energy Rev 82(2):2843–2868

Chapter 4
PCMs in Building Structure

PCMs incorporated directly into the building structure such as walls, ceilings, and floors are microencapsulated [1]. Commonly used building materials with microencapsulated PCMs are gypsum plasterboards [2], plaster [3, 4], concrete [5], brick [6, 7], façade [8], and glazing systems [9]. The basic concept for interior temperature control using PCM integrated into the building structure and glazing is shown in Fig. 4.1.

Referring to the figure, excess heat from the room is stored in the PCM by lowering room temperature during a hot day. Stored heat is removed by active or passive ventilation at night. Microencapsulated PCMs are usually incorporated in walls structure while macroencapsulated PCMs can be incorporated in floors and ceilings.

Different researchers employ different methods that have been conducted on thermo-physical properties of new phase change materials. They applied theoretical/analytical, experimental [10], and numerical studies on container designs and integration of the PCM in building components. Tables 4.1 and 4.2 summarize experimental and numerical studies conducted on PCM applications in buildings and different geometries, respectively.

As it was shown in tables, different shapes of the container were studied (spherical, cylindrical, rectangular, or encapsulated PCMs), and various discretization methods were applied as well (FEM, FDM, FVM). To improve the thermal mass of the building, a number of experimental studies were conducted. Different parts of buildings were considered as PCM containers (gypsum boards, PVC panels, fiber panels, sandwich panels, tubes, air-conditioning ducts, etc.), which were tested in the laboratory or outdoor conditions.

Many studies have included the numerical analysis of PCM-integrated elements in a building. Table 4.3 shows a brief overview among different PCM-incorporated building elements that exhibit in some way the impact PCM has on the surrounding environment, be it interior or exterior.

© Springer Nature Switzerland AG 2020

B. Duraković, *PCM-Based Building Envelope Systems*, Green Energy and Technology, https://doi.org/10.1007/978-3-030-38335-0_4

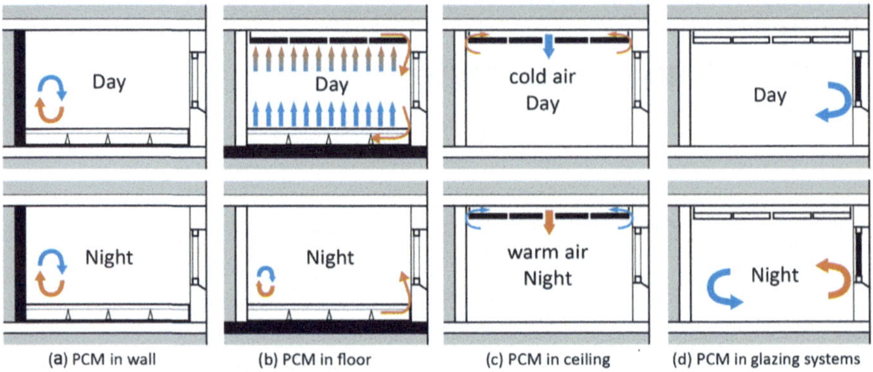

Fig. 4.1 PCM in building structure and glazing

Table 4.1 Exp. studies on PCM containers in building applications

Commercial PCM	Container	Container size (m)	References
RT27	Glazed system	0.3 × 0.4 × 0.02	[11, 12]
Butyl stearate	Gypsum	2.88 × 2.22 × 2.24	[2, 13]
$CaCl_2 \cdot 6H_2O$	Tube	2.95 × 4.43 × 2	[14]
Capric–lauric non-eutectic	Gypsum	5 × 3.3 × 2.8	[15]
ERMEST2325	Gypsum	2.27 × 2.29 × 2.45	[2, 16, 17]
Eutectic salt	Sandwich panel	4.37 × 3.39 × 2.7	[18]
MICRONAL26	Gypsum	2.4 × 2.4 × 2.4	[5, 19]
RT20	Gypsum	0.7 × 0.7 × 0.7	[20, 21]
RT25	Fiber panel	2.4 × 2.4 × 2.4	[6]
RT27	Fiber panel	2.4 × 2.4 × 2.4	[19]
Paraffin	Shape stabilized	0.575 × 0.453 × 0.463	[22, 23]
PEG600	PVC panel	0.9 × 0.9 × 0.9	[24]
PC5001; PCM5008	Aluminum foils	2.7 × 2 × 1.5	[25]
U1	Tube	1.83 × 1.83 × 1.22	[26]
U2	Steel	1.22 × 1.22 × 1.44	[27]
U3	Gypsum	2.95 × 4.43 × 2	[14]
U4	Energain	3.1 × 3.1 × 2.5	[3, 4, 28]
SP25A8	Fiber panel	2.4 m × 2.4 m × 2.4 m	[6]

Table 4.2 Computational studies on different PCM geometries

Commercial PCM	Container geometry	Dimension	Discr. meth.[a]	Validation
Gallium	Rectangular	2D(*X*, *Y*)	FDM	Experimental
Paraffin RT27	Rectangular	2D(*X*, *Y*)	FVM	Experimental and numerical [11, 12, 29]
Paraffin wax RII-56	Bags in duct	3D(*X*, *Y*, *Z*)	FDM	Experimental and numerical [30]
Paraffin RT30	Cylindrical	2D(*R*, *X*)[b]	FVM	Experimental
n-Octadecane	Cylindrical	2D(*R*, *X*)	FDM	Experimental
4 PCM	Cylindrical	2D(*R*, *X*)	FDM	–
RT60	Cylindrical	2D(*R*, *U*)	FDM	Experimental
Paraffin130/135 Type1	Cylindrical	2D(*R*, *U*)	FDM	Experimental
Paraffin wax	Cylindrical	2D(*R*, *U*)	FDM	Experimental
n-Hexacosane	Cylindrical	1D(*R*)	–	Experimental
4PCM	Cylindrical	2D(*R*, *X*)	FDM	–
2PCM	Cylindrical	2D(*R*, *X*)	FDM	[31]
5PCM	Cylindrical	2D(*R*, *X*)	FEM	–
3PCM	Cylindrical	2D(*R*, *X*)	FDM	[32]
Water/ice	Spherical	1D(*R*)	FDM	Experimental
Water glycol mixtures	Spherical	1D(*R*)	FDM	Experimental

[a]FDM—Finite difference method; FEM—Finite elements method; FVM—Finite volume method
[b]*R*, *X*; *R*, *U*; *R*—represents cylindrical coordinate system (cylindrical geometry)

Table 4.3 Comparison among studies on PCM effectiveness

Building element	Climatic region	Type of experiment	Results	Notes	References
Wallboard	Beijing, China		Reduction in energy consumption more than 10% during winter	Study limited to four days	[33]
Wall	Madrid, Spain		Reduction in maximum and amplitude of heat flux	6 days	[34]

(continued)

Table 4.3 (continued)

Building element	Climatic region	Type of experiment	Results	Notes	References
Building envelope	Six Australian climatic zones	Numerical	17–23% annual energy savings	PCM can decline the energy consumption under cold, mild, and warm temperature climates. However, the integration of PCM in buildings under hot and humid climate has very limited impact on the energy consumption	[35]
LHSU	Five Chinese climatic zones	Numerical and experimental	Energy-savings potential of the latent heat storage unit was 50%	LHSU: a technology that combines phase change materials (PCMs) with a natural cold source is proposed to reduce the space cooling energy of telecommunications base stations	[36]
Building envelope	Kut, Iraq	Experimental	Indoor temperature of reduced by 2.18 °C, cooling load of the zone for peak hour in day decreased by 20.9%	PCM sheets with 1 cm thickness on all envelopes were used	[37]
Building envelope	Five climatic regions of Iran	Numerical	Reduction in energy demand was 17.5% in warm/dry climate, 10.4% and in mild/semi-arid climate. Cooling energy consumption decreased around 12.3% in cold climate and 9.8% in mild/humid climate	Double PCM inner layer. The double layer PCM system can significantly decrease the radiation heat loss from human body on cold days	[38]

(continued)

Table 4.3 (continued)

Building element	Climatic region	Type of experiment	Results	Notes	References
Night ventilation	Hot-arid climate		A 45.5% drop of the annual cooling load	With the exception of the ground floor, where the use of PCMs augmented the overall cooling load	[39]
Facade			Ventilating facades with fins filled with PCMs and PCMs in hollow core slabs is superior than the prevalent usage of PCM both for cooling power and storage capacity		[40]
Hybrid cooling system	Beijing		SSPCM plates integrated with night ventilation in comparison with the case without SSPCM and NV could save around 76% of daytime cooling load demand		[41]
Floor		Experimental	Provided 89% of daily cooling demand. Daytime AC was limited to around 3 h per day from 1 p.m.	In conjunction with night ventilation	[42]
Brick wall			Flux at the indoor space can be reduced by 17.55% when three brick cylinders filled with PCM and placed at the centerline of the bricks		[43]

4.1 Application in Walls

Application of PCM in wall structure has been studied for a longer period as interior temperature control mechanism. The basic concept shown in Fig. 4.1a is suitable for passive cooling/heating in which the heat from the warm air over the day is removed by the PCM. At night, stored heat is removed to the exterior (Fig. 4.2).

4.1.1 Application in Bricks

Alawadhi [43] studied thermal analysis of a brick wall containing phase change material by analyzing PCM's quantity, type, and location in the brick. Results showed that the heat flux at the indoor space can be reduced by 17.55% when three brick cylinders are filled with PCM and placed at the centerline of the bricks.

Ning et al. [44] experimentally examined the impact of PCM application in the building envelope on interior temperature fluctuations in a coastal city during the summer season. The setup is shown in Fig. 4.3. Chamber is equipped with heat and moisture load generating units. Two adjacent rooms A and B simulated different environmental conditions [44].

The return air temperature variation within the framework of one day can be read from Fig. 4.4. It is notable that both cases seem to almost be the same from 0:00 to 5:00, but from 7:00 to 23:00 case 2 remains lower values than the case 1 [44].

The research has also come to the conclusion that increasing the melting temperature can reduce the variation of return air temperature at the period of high temperature [44].

Implementation of light thermal mass walls in a Mediterranean climate with the help of PCM was investigated by Faraji [45]. The aim was to check the feasibility

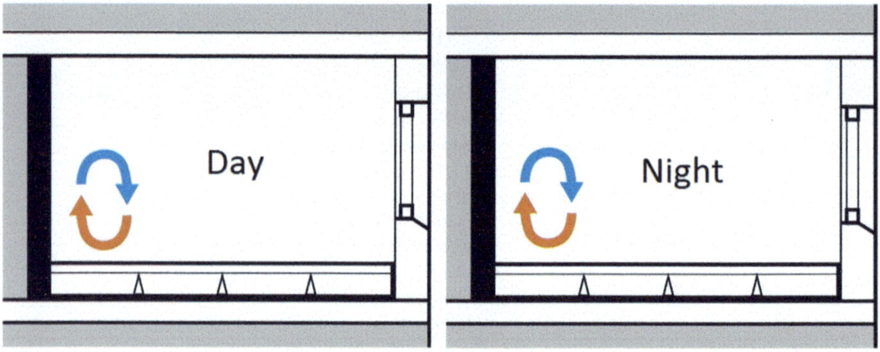

Fig. 4.2 PCM in building wall structure

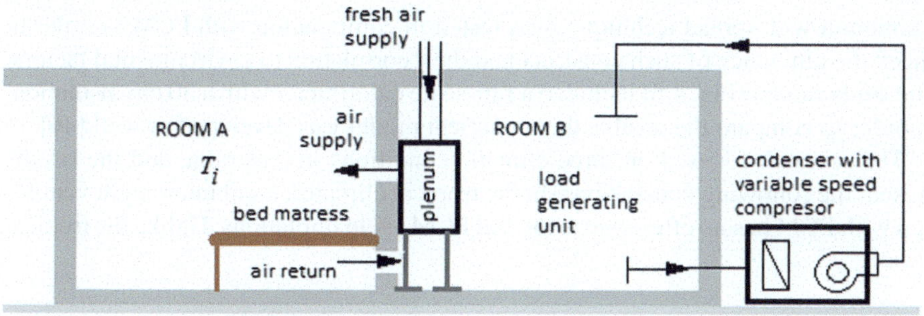

Fig. 4.3 Environmental chamber

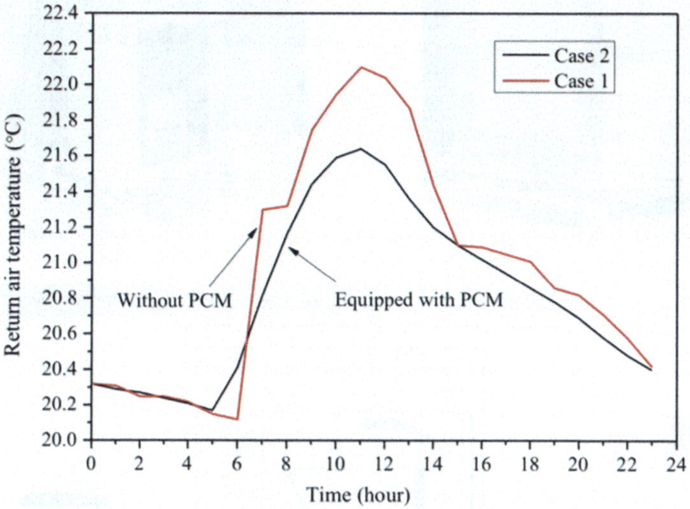

Fig. 4.4 Return air temperature for PCM equipped and non-equipped case [44] [This article was published in Energy Procedia, 142, Ning, Mao; Jingyu, Hao; Dongmei, Pan; Shengchun, Liu; Mengjie, Song, Investigations on thermal environment in residential buildings with PCM embedded in external wall, 1888–1895, Copyright The Authors (2017)]

of maintaining similar comfort levels throughout the year. The model of a PCM-incorporated south wall was used in the experiment under the typical working conditions. The results bear the following conclusions: The inner temperature swings remarkably decrease with the use of PCM, the concrete/PCM composite wall performs well in above-mentioned conditions, subcooling and subheating occur in cases when solidification and melting processes are not finished and further investigation is required, and special care should be taken in the selection of PCM [45].

The investigations have furthermore elaborated on techniques and ways in which PCM can be utilized with the emergence of lightweight structures. Such an example can be found in a study [46] conducted in Australia where the night ventilation, a common widespread technique, was tested in combination with PCM in order to check the efficiency of such construct and their correlation to environmental factors. The study model (Fig. 4.5) included a full-scale calorimeter with a 30 cm insulation. In order to compare the results, the numerical model was developed as well [46].

The research focused on three climatic conditions in Australia, and the results present the following conclusions. In the tropical climates, combining night ventilation with PCM has no effect due to the full PCM cycle limitations. Unlike the tropical

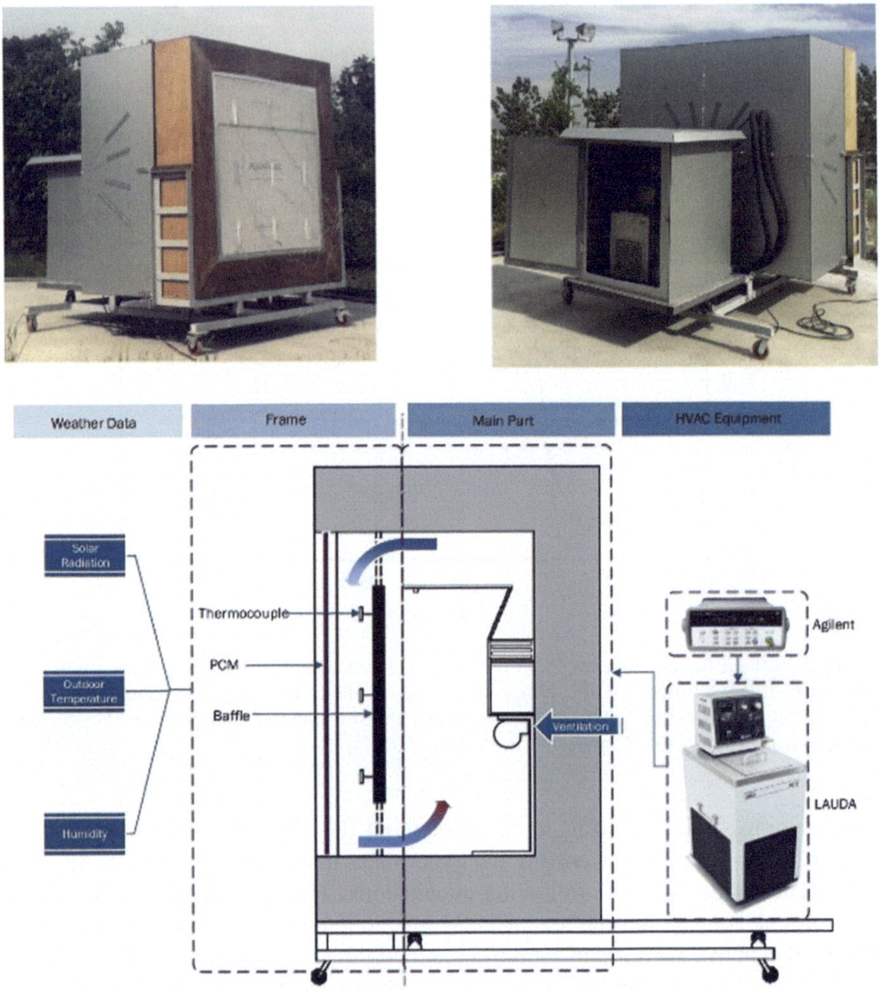

Fig. 4.5 Calorimeter scheme [46] [This article was published in Building and Environment, 147, Solgi, Ebrahim; Hamedani, Zahra; Fernando, Ruwan; Kari, Behrouz Mohammad; Skates, Henry, A parametric study of phase change material behaviour when used with night ventilation in different climatic zones, 327–336, Copyright Elsevier (2019)]

climate, subtropical and hot–dry climates foster climatic opportunities to implement this hybrid system, where the selection of cooling thermostat set points and thermal insulation affect the optimal PCM temperature. However, the temperature is not dependent on the PCM thickness. Nonetheless, thickening raises energy saving [46].

Navarro et al. studied commercially available PCMs for incorporation into a building concrete matrix. The results suggest that materials in bulk have higher thermal conductivity and specific heat values over micro and macroencapsulated materials. Despite their good performance, bulk materials are more difficult to impregnate into a concrete matrix. Thus, microencapsulated PCM was chosen to be the best fit for the purpose of this study [47].

Derradji et al. conducted a numerical study comparing office rooms with and without PCM-incorporated walls. The study was situated in Algerian climate measuring 3.5 m length, 3 m wide, and 3 m high room, with external south-facing wall having a double-glazed window, and other walls are concrete interior walls. The roof is 15 cm concrete complemented with 5 cm EPS [48].

Results show that without PCM, temperatures in winter vary from 16 to 21 °C, whereas with PCM, temperatures vary from 18 to 22 °C [48] (Fig. 4.6).

In the summertime period, temperatures vary from 33 to 37 °C without PCM, which is considerably higher than the outside temperatures between 20 and 31 °C. In the case of PCM, incorporated office temperatures are significantly lower, between 25 and 28 °C [48].

The study did not include insulation except on the roof. The use of PCM in the concrete ceiling and hollow brick walls affected winter temperature fluctuations by 4 °C and by 7 °C summer temperatures. Presence of 30% of PCM in gypsum panels in the interior surfaces makes 25% savings for both cooling and heating energy demands [48] (Fig. 4.7).

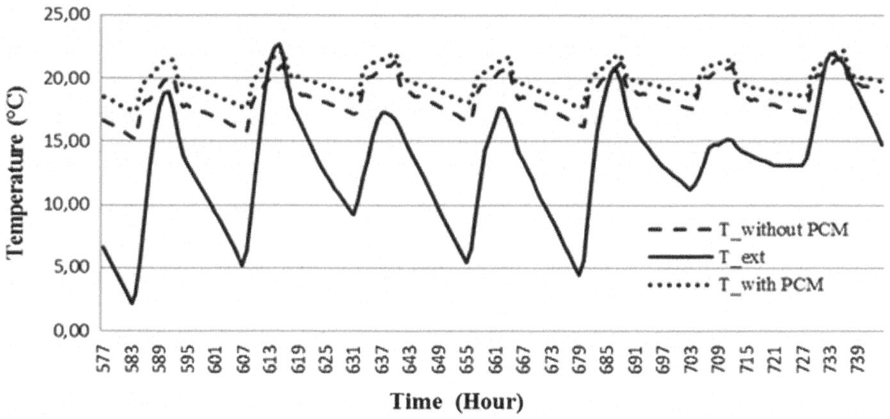

Fig. 4.6 Ambient temperatures in the period: January 25–31 [48] (This article was published in Energy Procedia, 107, Derradji, Lotfi; Errebai, Farid Boudali; Amara, Mohamed, Effect of PCM in Improving the Thermal Comfort in Buildings, 157–161, Copyright The Authors (2017)).

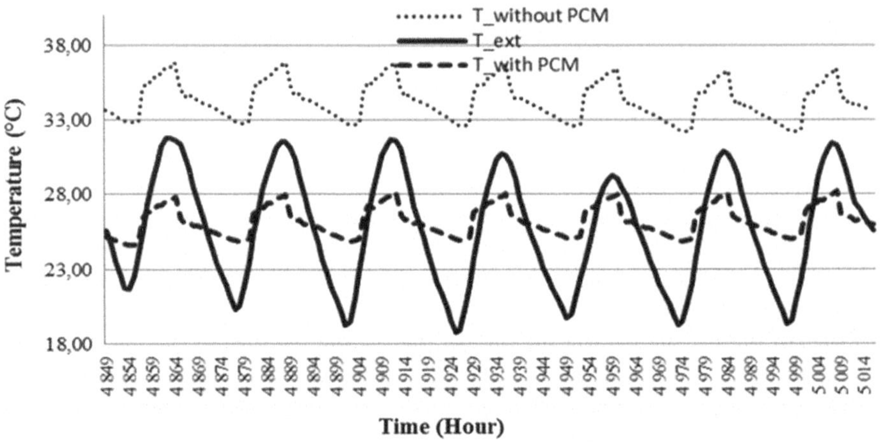

Fig. 4.7 Ambient temperatures in the period: July 25–31 (This article was published in Energy Procedia, 107, Derradji, Lotfi; Errebai, Farid Boudali; Amara, Mohamed, Effect of PCM in Improving the Thermal Comfort in Buildings, 157–161, Copyright The Authors (2017)). [48]

Panayiotou et al. theoretically evaluated the effects of application of PCM in a typical dwelling unit in Cyprus. They did experiments with test cubicle dimensioned $W \times L \times H = 3 \times 2 \times 3$ m, facing to east–west and north–south. The climate condition is Mediterranean having cooling and heating demands throughout the year [49].

Application of macroencapsulated PCM layer in combination with brick was done in three different positions: double brick wall with a PCM interlayer (new construction), PCM placed inside behind plaster (new and old construction), and PCM placed outside behind plaster (new and old construction). A PCM layer was also applied in the inner layer of the roof, behind the plaster. PCM used in the experiment was BioPCmat™ M91 from Phase Change Energy Solutions Company, with a melting temperature at 29 °C [49].

Results of various case simulations show that most energy savings (28.6%, 118.5 kWh/year/m^2) are achieved when PCM is in the outer layer. Energy savings are greater in cooling than heating due to the higher outside temperatures than the PCM melting temperature which is 29 °C. This supports the melting/solidification cycle to be complete. In the wintertime, PCM absorbed a part of the heat from heating devices and from incident solar radiation, which PCM released later. For the summertime, the best option is insulation [50] combined with PCM; whereas, for the wintertime it is only the insulation. Ultimately the combined version is the best choice since benefits of both layers are utilized in all weather conditions [49].

Life cycle cost of a typical Cypriot dwelling of total area 133 m^2, with four inhabitants, and no thermal insulation, consisted of three bedrooms, kitchen, living room, bathroom, and dining room was calculated. The cost of optimum PCM case was €22,490 with payback period 14½ years. The cost of the combined case was €23,259 with payback period 7½ years [49].

Hichem et al. made a numerical and experimental study on the optimal position and quantity of PCM in a brick wall in Algeria, Ouargla. The wall is built with bricks (30 × 20 × 15 cm) of 12 cavities measuring 4 × 3.667 cm [51] (Fig. 4.8).

The experimental setup is shown on Fig. 4.9. In the study, the internal temperature was imposed to be 27 °C. All surfaces except for outer and inner were insulated. A glass was placed on top of the brick in order to hinder heat transfer via convection with the external environment. Five types of PCMs with different melting temperatures (29.9–52 °C) in three positions are outer, middle, and inner [51] (Figs. 4.10 and 4.11).

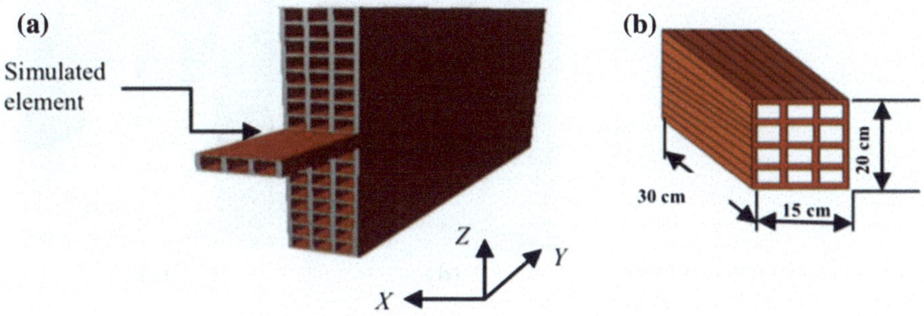

Fig. 4.8 **a** External wall of the building, **b** a typical hollow brick with twelve recesses set into three columns and four rows [51] [This article was published in Energy Procedia, 36, Hichem, Necib; Noureddine, Settou; Nadia, Saifi; Djamila, Damene, Experimental and numerical study of a usual brick filled with PCM to improve the thermal inertia of buildings, 766–775, Copyright The Authors (2013)]

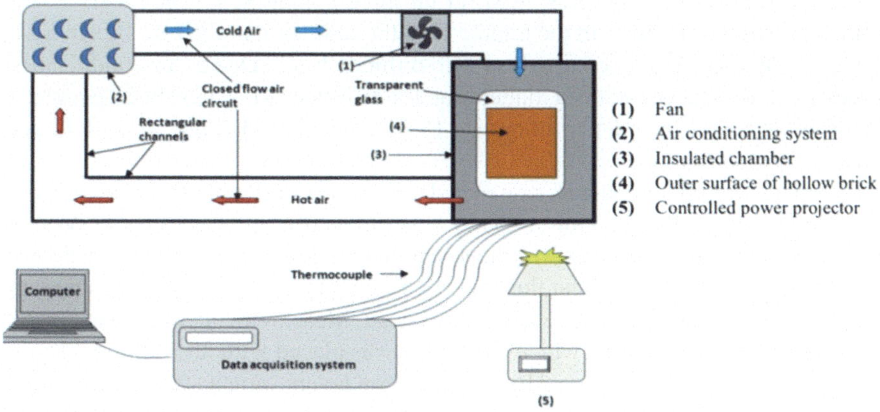

Fig. 4.9 Experimental setup [51] [This article was published in Energy Procedia, 36, Hichem, Necib; Noureddine, Settou; Nadia, Saifi; Djamila, Damene, Experimental and numerical study of a usual brick filled with PCM to improve the thermal inertia of buildings, 766–775, Copyright The Authors (2013)]

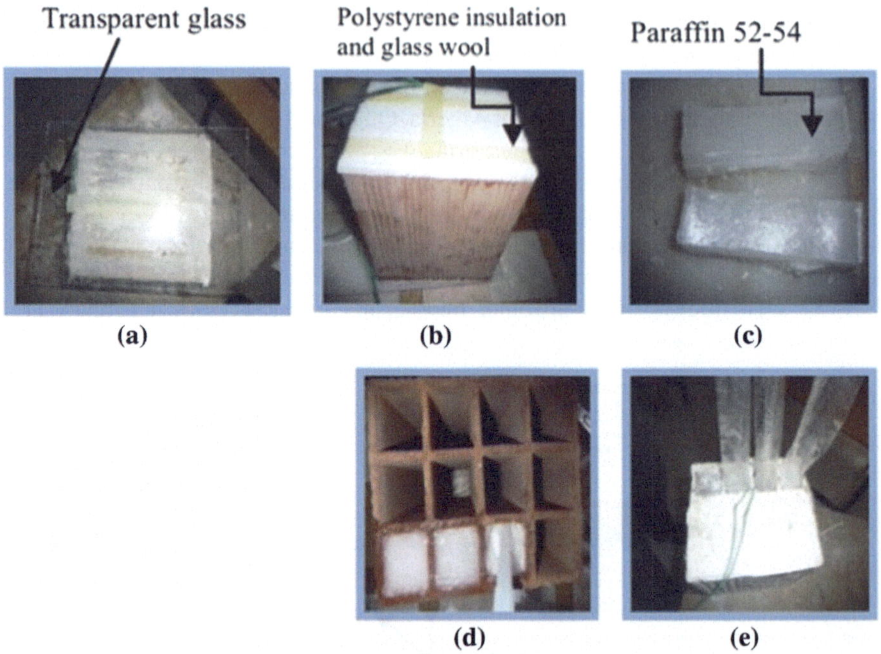

Fig. 4.10 Materials with PCM, **a** glass , **b** EPS and glass wool insulation, **c** paraffin 52–54, **d** hollow brick, **e** transparent plastic film [51] [This article was published in Energy Procedia, 36, Hichem, Necib; Noureddine, Settou; Nadia, Saifi; Djamila, Damene, Experimental and numerical study of a usual brick filled with PCM to improve the thermal inertia of buildings, 766–775, Copyright The Authors (2013)]

Results suggest that the best PCM, in all three positions, is $CaCl_2{\cdot}6H_2O$. The optimal position of PCM is in the middle with the reduction in the heat flux in a 24-hour cycle up to 82.1%, in comparison with ordinary brick. Due to the solidification process, a study that considers a longer time frame is needed. A new brick geometry should be invented in order to minimize PCM amount [51] (Figs. 4.11, 4.12 and 4.13).

Zastawna-Rumin and Nowak researched on benefits of double PCM layer application in Polish weather conditions. They experimentally analyzed three PCM configurations. In the first two stages of the research, individual layers were investigated in terms of temperature and heat flux density. Based on the compliance of materials, combinations of them were made in the third stage. Measurements were made inside of two chambers, in between which a partition was placed with all additional layers mounted on a frame (195 cm × 210 cm). Partition layers were a styrofoam (15 cm) and a drywall. PCMs were chosen as such that in case of great thermal load, and the second layer gets activated [52].

Results suggest that bio PCM mats used as the second layer are most activated but utilized not over 50% of its potential. In the case of double layer system, the cooling phase requires greater length and intensity. A cardboard containing PCM 23 and bio

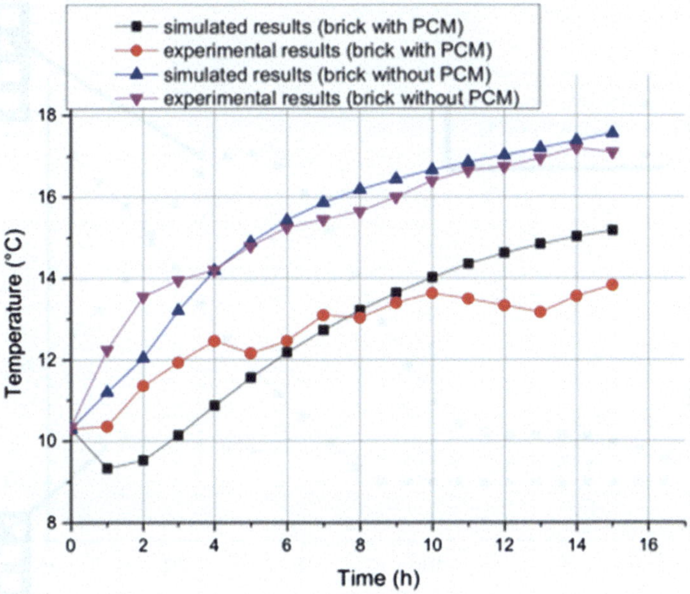

Fig. 4.11 Comparison between numerical and experimental data results of the average interior surface temperature [51] [This article was published in Energy Procedia, 36, Hichem, Necib; Noureddine, Settou; Nadia, Saifi; Djamila, Damene, Experimental and numerical study of a usual brick filled with PCM to improve the thermal inertia of buildings, 766–775, Copyright The Authors (2013)]

Fig. 4.12 Total heat flux reduction through the brick [51] [This article was published in Energy Procedia, 36, Hichem, Necib; Noureddine, Settou; Nadia, Saifi; Djamila, Damene, Experimental and numerical study of a usual brick filled with PCM to improve the thermal inertia of buildings, 766–775, Copyright The Authors (2013)]

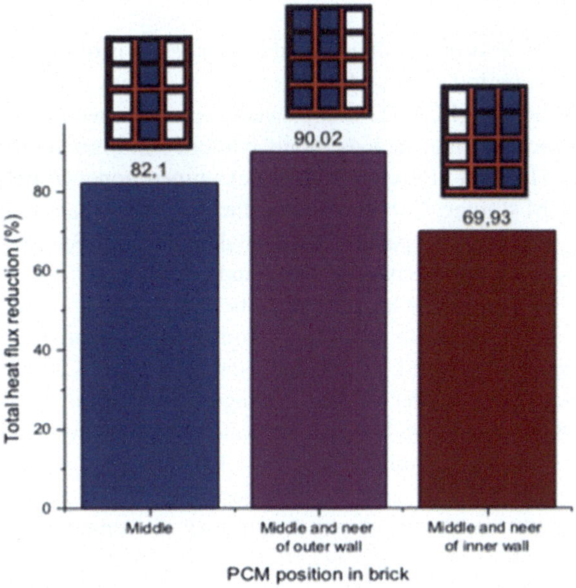

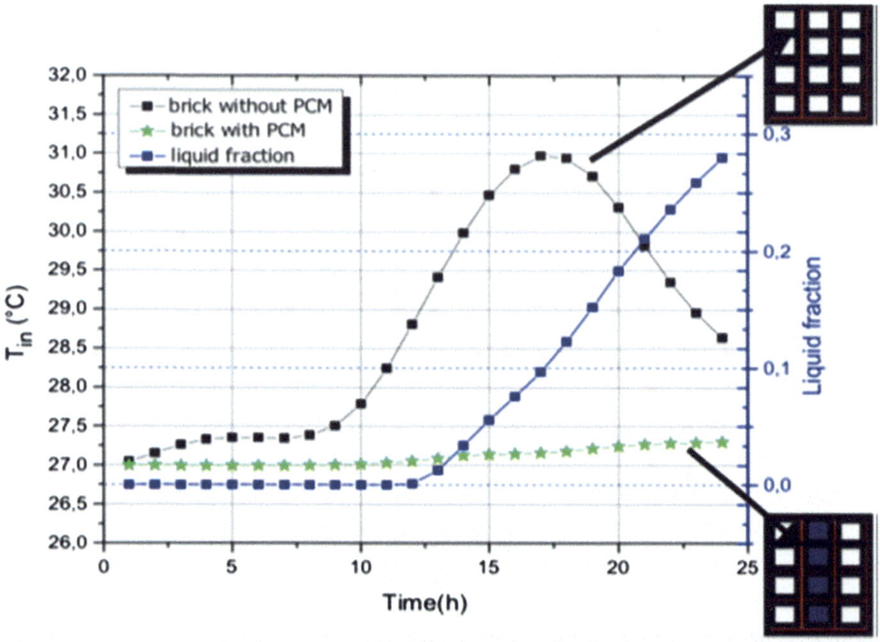

Fig. 4.13 Variation of liquid fraction and average temperature of interior brick surface [51] [This article was published in Energy Procedia, 36, Hichem, Necib; Noureddine, Settou; Nadia, Saifi; Djamila, Damene, Experimental and numerical study of a usual brick filled with PCM to improve the thermal inertia of buildings, 766–775, Copyright The Authors (2013)]

PCM 35 is most efficient, but its efficiency is most noticeable in large heat loads in a short period of time. Such example is a room with a window. Combination of PCM 26 and bio PCM 23 mat proved to perform well in smaller thermal loads and long period of time, such as rooms without windows and/or ventilation [52].

Hasse et al. experimentally examined honeycomb panels as a means of paraffin PCM storage. Simultaneously, they tested the same sample filled with water and air in order to compare their thermal responses. For low-temperature conditions, best PCMs to use are paraffins, salt and salt hydrates, and fatty acids. Such materials with low thermal conductivity once heated can become liquid at some parts and remain solid at other. In order to obtain the same consistency, a high conductive material is introduced. In the study, encapsulated PCM with 27 °C melting temperature was used in order to avoid leakage. It is a commercially available PCM, LINPAR® 1820. Aluminum honeycomb panels were chosen for its enhancement property and containment ability [53] (Figs. 4.15 and 4.16).

Honeycomb panels, 2 cm deep, have 6 mm cells with 7 μm cell wall thickness. Panels were covered with 1 mm thick aluminum sheets [53] (Fig. 4.16).

From Fig. 4.16, it becomes obvious that honeycomb arranged PCM is able to store around three times more energy than the water sample. The experiment was numerically validated. Authors stressed that fins immersed in PCM improve heat distribution and called for further research on this novel methodology of heat transfer through the building envelope [53].

Thantong and Chantawong experimentally compared the performance of two small houses, in order to assess how solar collector in combination with PCM can contribute to the natural ventilation. The experiment was conducted under Thailand's weather conditions which are humid tropical climate with average temperatures around 20–30 °C and average humidity around 60% [54].

Fig. 4.14 Honeycomb panel with paraffin [53] [This article was published in Energy and Buildings, 43, Hasse, Colas; Grenet, Manuel; Bontemps, André; Dendievel, Rémy; Sallée, Hébert, Realization, test and modelling of honeycomb wallboards containing a Phase Change Material, 232–238, Copyright Elsevier (2011)]

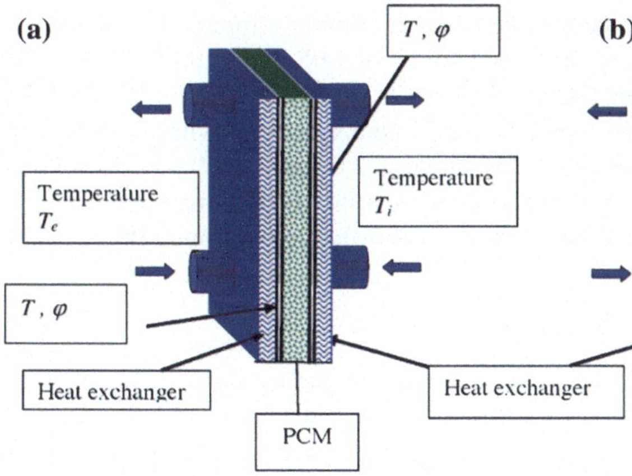

Fig. 4.15 Experimental setup [53] [This article was published in Energy and Buildings, 43, Hasse, Colas; Grenet, Manuel; Bontemps, André; Dendievel, Rémy; Sallée, Hébert, Realization, test and modelling of honeycomb wallboards containing a Phase Change Material, 232–238, Copyright Elsevier (2011)]

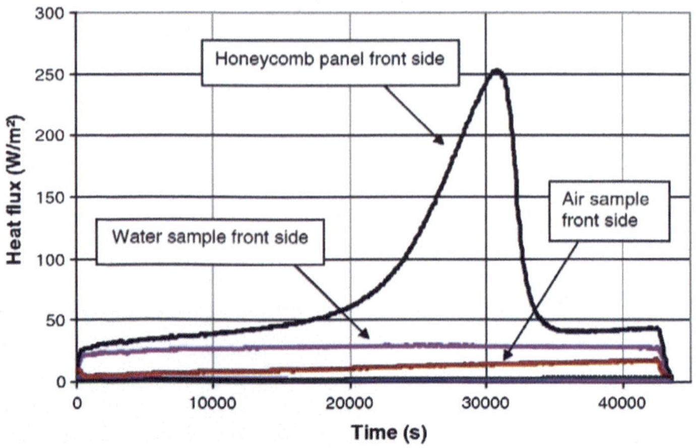

Fig. 4.16 Heat fluxes of three samples [53] [This article was published in Energy and Buildings, 43, Hasse, Colas; Grenet, Manuel; Bontemps, André; Dendievel, Rémy; Sallée, Hébert, Realization, test and modelling of honeycomb wallboards containing a Phase Change Material, 232–238, Copyright Elsevier (2011)]

The solar wall collector is a double wall with an air gap that has openings on upper and lower side. The air gets released to the outside from the upper opening, and the interior air is discharged from the lower opening into the wall chimney [54].

The outer wall is consisted of black cement board (as an absorber), PCM which is a paraffin wax (as an accumulator), and a zinc sheet (for transferring heat to the air gap). The inner wall is AAC wall that has good thermal protection properties. Once the air gap temperature gets higher than the indoor temperature, natural ventilation occurs in the wall. The experimental house, of 4.05 m^3 volume, is a single room with AAC walls and red concrete tile roof with an insulated attic. House 1 had a solar wall collector with PCM, while house 2 had a solar wall without PCM [54].

Heat flux results show that PCM-integrated house had 59.63% lower heat transfer through the south wall. Results also suggest that PCM-integrated house had better air circulation of the ventilation with faster velocity by around 96.62% [54].

Performance of passive cooling using microencapsulated PCM as part of wall structure has been studied by different authors [5, 7, 13, 19, 38, 44, 45]. Results show increased thermal performances of the wall by reducing thermal amplitude, as well as increasing the time delay. Most of the available research work is concerned with the evaluation of different kinds of wallboards [4, 15, 16, 55]. Studies included experimental and numerical methods in assessing thermal performances of wallboards. Results show that the board can contribute in reducing temperature fluctuations in the interior.

4.1.2 Application in Wallboards

Ahangari and Maerefat [38] numerically investigated the behavior of double layer PCM wallboard system in five climatic regions of Iran. The model was composed of a south-facing wall with a double-glazed window and a north-facing wall with a wooden door. The simulation did not take into consideration heat transfer in the interior surfaces [38].

The building envelope was designed according to the common local materials where the first layer of PCM was regulating thermal comfort in cold months and second layer of PCM regulated the hot months, since the melting temperature of the second layer is higher [38].

The results indicate that the interior orientation of PCM layers is the best performing. With such disposition, the melting points of PCM are close to average room temperature resulting in better thermal comfort conditions.

In moderate and semi-arid climate, thermal comfort changes from 63 to 75%, and in arid climate from 73 to 93%. When it comes to hot and cold climates of Iran, PCM has a neglectful impact, and in moderate and humid climate PCM improves the thermal comfort from 14 to 19%. The results support the argument that in arid climates, temperature differences between the daytime and nighttime facilitate the completion of the phase transition cycle. Moreover, the heating energy consumption is reduced by 10.4% in semi-arid and by 17.5% in dry climate [38].

The simulation included five variations of PCM-incorporated exterior surfaces:

South wall,
North and south walls,
North and south walls and ceiling,
North, south, and east walls and ceiling,
All exterior surfaces except floor.

The reduction in energy consumption is between 6 and 10%. Increasing the number of exterior double PCM layer incorporated envelopes directly contributes to the reduction in energy consumption. The greatest efficacy can be found in the south-oriented wall [38].

Authors concluded that the desired melting point of the first layer should be 1 °C lower than the optimal indoor temperature in winter, and in the case of second layer should be 2–3 °C above the summer interior temperature. Considering 22 and 24 °C as room air temperature in winter and summer, the optimal melting points are 21 and 26–27 °C during winter and summer, respectively. The double layer system is best suit for warm/dry climate and mild/semi-arid climate according to the results. Findings also suggest that the double layer PCM system can significantly decrease the radiation heat loss from human body during cold days [38].

Chhugani et al. compared two rooms equipped with two types of PCM wallboards, Knauf Comfortboard-23® and DuPont Energain® Board, in terms of their thermal performance and measured them against a typical office room without PCM. The experiment was conducted in Bayern, Germany, with moderate climate [56].

Knauf Comfortboard-23® is 12.5 mm thick with 200 kJ/m^2 heat storage capacity, melting enthalpy between 18 and 23 °C, and contains 80% gypsum and 20% microencapsulated paraffin.

DuPont Energain® Board is 5.26 mm thick with 515 kJ/m^2 heat storage capacity, melting enthalpy between 17 and 21 °C, and contains 60% paraffin. Since paraffin is flammable, a fire-resistant coating was applied during the testing [56].

Rooms R111, R112, and R113 contained PCMs with PCM cooling ceilings, while R110 contained no PCM with a conventional cooling ceiling and was used as a reference room. Each PCM board was 16.90 m^2. The reference room had a gypsum board mounted on a concrete wall on the east and gypsum wall on the west. DuPont Energain® Boards were additionally covered by gypsum boards for fire protection [56].

The experiment was conducted with no users involved and shut blinds. Instead, artificial heat was produced by electrical heaters that were heating from 8:30 to 20:30. Every room had five of them making 858 W in total. Before the experiment started, wall temperature was 20 °C. The outside temperature was 26 °C. The maximum rise of the temperature during the experiment was 26 °C [56].

Negative heat flux (passive room cooling) is a blue-colored background, and positive heat flux (regeneration of PCM wallboards) is the red-colored background. The red arrow shows that PCM melted and beyond that point passive cooling power decreased [56].

Results show that PCM wallboards can store twice the heat in comparison with conventional gypsum boards. It is stressed on the fact that proper boards should be chosen for efficient regeneration behavior. While DuPoint board's average regeneration rate was nearly 1%, Knauf Comfortboard-23 had below 20% in the summer months [56].

4.1.3 Application in Façade Elements

The effectiveness of PCM in energy conservation has been proven by many researchers to be efficient and suitable, provided that certain parameters are fulfilled in accordance with the environment. In order to complete the study on PCM application possibilities, façade as the skin of the building has to be considered. PCM has been proven to be well performing in the wall structure, thus it is not surprising to find out that such trends can also be recognized in the façade implementations.

A model of integrated double skin facade (DSF) and phase change material (PCM) blind system have been developed and analyzed by Li et al. [57]. The study proposed an integrated microencapsulated composite laminated PSM blind system [57].

The model has a blind structure consisting laminated composite PCM blades having a multilayered blade structure. Slats are tilted at the angle of 30°. The mechanism assumes that once the cavity temperature exceeds the melting point of PCM, it can absorb the solar heat. Later on, the stored heat will be discharged by ventilation once the temperature in the cavity drops. The study was compared to the conventional aluminum blinds [57].

Study indicated that the PCM case was reducing the cavity air temperature more than the conventional aluminum blind. Such indications imply that PCM blind can store more heat during the melting phase instead of releasing it into the cavity during the highest temperature intervals. Authors concluded that PCM integration in the façade of the building has a huge potential in the future implementations [57].

4.2 Application in Floors

Application of PCM in floor is shown in Fig. 4.17. The concept of cooling in daytime hours is achieved by extracting the warm air from the room from the top of the room and bringing fresh air inside under floorboards. The air is cooled by melting the PCM in the floorboards. At night, cold night air can be circulated under the floor space to cool down the PCM and remove the stored heat. This concept requires active system as well. Some experimental and numerical studies of the thermal performances of floor-based PCM were done including floor supply air-conditioning system [1] and floor heating [22, 23].

The case of PCM-based floor complimenting an air-conditioning system has been simulated in software environment and measured against chilled ceiling system. Cheng et al. took Chinese weather conditions and applied them to a typical room in an office building.

Two building models were simulated in the most demanding months for cooling from June to September. The first model included chilled ceiling system; whereas, the second model in addition included a PCM layer in the floor [58].

The results indicate that PCM has a significant effect on stabilizing the oscillations in the heat transfer between the floor surface and the air. The PCM layer showed its

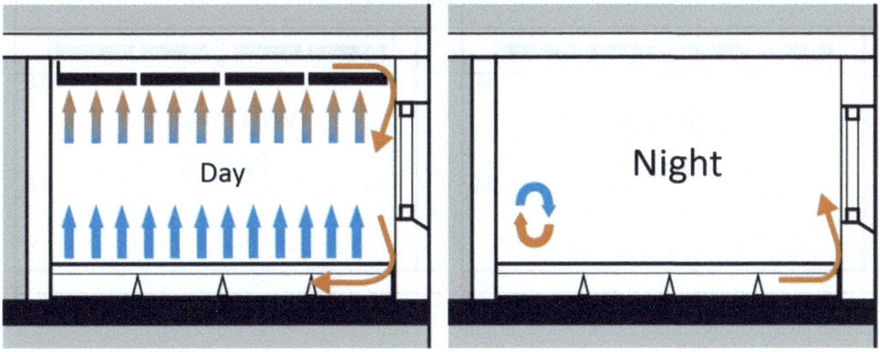

Fig. 4.17 PCM-incorporated floor

ability to store large amounts of thermal energy during the cold periods and release it when the air gets hotter reducing peak cooling loads during summer [58].

4.3 Application in Ceilings

Application of PCM in ceiling is based on using forced convection for storing the heat in the suspended ceiling panels shown in Fig. 4.18. This principle can be developed into an active system by providing a small fan to blow air over the PCM panels to foster heat removal at night [59]. Warm air from the room is forced to move across the PCM located in suspended ceiling channel during the day. The PCM cools the air by absorbing the heat. At night, the stored heat is removed by circulating cool air to the outside of the building. To speed up heat transfer process between the air and the PCM, aluminum panels with cooling ribs have been developed [59].

Integrating PCM ceiling panels and its implications on energy consumption savings has been investigated by Yahaya and Ahmad [60]. Single space house fostered the comparison between an ordinary ceiling and the PCM ceiling to evaluate the effectiveness of both systems in terms of passive cooling design. The assessment was based on indoor air temperature reduction and cooling load for the comfort temperature of 25 °C. The PCM that could be obtained from the palm oil, produced by the immersion process of gypsum board in a eutectic mixture of lauric-stearic acid, was employed. This mixture is non-corrosive, non-flammable, and thus more preferred to be adopted in building structures. Its higher melting point and latent heat capacity make it suitable for ceiling applications in order to reduce downward heat flow in hot weather. The numerical model was predicting the indoor temperature taking into consideration the solar radiation. The study model was based in Subang Jaya, on the west coast of Peninsular Malaysia, in mid-January [60].

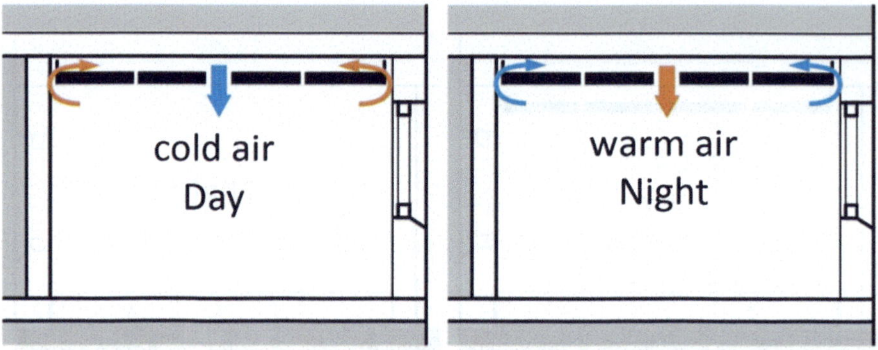

Fig. 4.18 PCM-incorporated ceiling or roofs

Pitched roof covered by concrete roof tiles with no insulation layer sits on top of a building envelop coated with high reflective paint. Results prove that energy efficiency is greatly improved by incorporating PCM panels in the ceiling [60].

Authors remind that further investigation needs to be conducted to prove the numerical model findings correct. Nonetheless, they recommend applying PCM as a ceiling component in order to improve single story building's thermal performance [60].

4.3.1 Application in Roof

Integration of PCM in all elements of the building envelope has been suggested throughout many researches. One of the least examined areas is the implementation of PCM into the roof structures. Until now, few studies have focused on the flat roof [61, 62]. The PCM ceiling has been tested and proven to be energy efficient. Further studies suggest that PCM roofs have the same contribution to the overall building envelope.

Hanchi et al. [62] compared two flat roofs from the local tradition of Casablanca and Ouarzazate. The study has been numerically confirmed the conventional flat roof as a reference to the second one with a specific sectional arrangement of materials. Similar to the wall applications, the case study roof has two PCMs to ensure thermal stability and comfort throughout the year [62].

The results indicate that the PCM flat roof is favorable in terms of energy conservation. Due to the changing climate and building renovation policies, heat demand may decrease, provided that a proper choice of PCM is made, coupled with an according room temperature selection [62].

4.4 Remarks

Passive building design with PCMs is a new and as well a very active research area. Even though a considerable amount of literature has been published on the topic of PCM, this material is fairly new to the market and the building construction practices. Until now, we had a chance to explore its application though selected study cases and many are yet to emerge, inspired by the existing research that is never fully complete in its detail and conclusions. Through this book, it can be easily understood that PCM has a huge potential in its application due to its reversible nature and lightweight properties. Moreover, its adjustability and widespread application and use will contribute to its ubiquity in future construction methods.

References

1. Soares N, Costa J, Gaspar A, Santos P (2013) Review of passive PCM latent heat thermal energy storage systems towards buildings' energy efficiency. Energy Build 59:82–103
2. Banu D, Feldman D, Hawes D (1998) Evaluation of thermal storage as latent heat in phase change material wallboard by differential scanning calorimetry and large scale thermal testing. Thermochim Acta 317(1):39–45
3. Liu H, Awbi H (2009) Performance of phase change material boards under natural convection. Build Environ 44(9):1788–1793
4. Kuznik F, Virgone J, Roux J-J (2008) Energetic efficiency of room wall containing PCM wallboard: a full-scale experimental investigation. Energy Build 40(2):148–156
5. Cabeza L, Castello´n C, Nogus M, Medrano M, Leppers R, Zubillaga O (2007) Use of microencapsulated PCM in concrete walls for energy savings. Energy Build 39(2):113–119
6. Castell A, Martorell I, Medrano M, Prez G, Cabeza L (2010) Experimental study of using PCM in brick constructive solutions for passive cooling. Energy Build 42(4):534–540
7. Bontemps A, Ahmad M, Johannès K, Sallée H (2011) Experimental and modelling study of twin cells with latent heat storage walls. Energy Build 43(9):2456–2461
8. Carbonari A, Fioretti R, Naticchia B, Principi P (2012) Experimental estimation of the solar properties of a switchable liquid shading system for glazed facades. Energy Build 45:299–310
9. Duraković B, Mešetović S (2019) Thermal performances of glazed energy storage systems with various storage materials: an experimental study. Sustain Cities Soc 45:422–430
10. Durakovic B (2017) Design of experiments application, concepts, examples: state of the art. Periodicals Eng Nat Sci 5(3):421–439
11. Duraković B, Torlak M (2017) Experimental and numerical study of a PCM window model as a thermal energy storage unit. Int J Low-Carbon Technol XII(3):272–280
12. Duraković B, Torlak M (2017) Simulation and experimental validation of phase change material and water used as heat storage medium in window applications. J Mat Environ Sci VIII(5):1837–1846 (2017). ISSN: 2028-2508. Copyright © 2017
13. Athienitis A, Liu C, Hawes D, Banu D, Feldman D (1997) Investigation of the thermal performance of a passive solar test-room with wall latent heat storage. Build Environ 32(5):405–410
14. Voelker C, Kornadt O, Ostry M (2008) Temperature reduction due to the application of phase change materials. Energy Build 40(5):937–944
15. Shileia L, Guohuib F, Nenga Z, Li D (2007) Experimental study and evaluation of latent heat storage in phase change materials wallboards. Energy Build 39(10):1088–1091
16. Scalata S, Banua D, Hawesa D, Parishb J, Haghighataa F, Feldman D (1996) Full scale thermal testing of latent heat storage in wallboard. Solar Energy Mat Solar Cells 44(1):49–61
17. Schossig P, Henning H-M, Gschwander S, Haussmann T (2005) Microencapsulated phase-change materials integrated into construction materials. Sol Energy Mater Sol Cells 89(2–3):297–306
18. Carbonari A, Grass M, Perna C, Principi P (2006) Numerical and experimental analyses of PCM containing sandwich panels for prefabricated walls. Energy Build 38(5):472–483
19. Castellón C, Castell A, Medrano M, Martorell I, Cabeza L (2009) Experimental study of PCM inclusion in different building envelopes. J Sol Energy Eng 131(4):041006
20. Fang X, Zhang Z (2006) A novel montmorillonite-based composite phase change material and its applications in thermal storage building materials. Energy Build 38(4):377–380

21. Fang X, Zhang Z, Chen Z (2008) Study on preparation of montmorillonite based composite phase change materials and their applications in thermal storage building materials. Energy Convers Manag 49(4):718–723
22. Lin K, Zhang Y, Xu X, Di H, Yang R, Qin P (2005) Experimental study of under-floor electric heating system with shape-stabilized PCM plates. Energy Build 37(3):215–220
23. Li J, Xue P, He H, Ding W, Han J (2009) Preparation and application effects of a novel form-stable phase change material as the thermal storage layer of an electric floor heating system. Energy Build 41(8):871–880
24. Ahmad M, Bontemps A, Sallée H, Quenard D (2006) Thermal testing and numerical simulation of a prototype cell using light wallboards coupling vacuum isolation panels and phase change material. Energy Build 38(6):673–681
25. Konuklu Y, Paksoy H (2009) Phase change material sandwich panels for managing solar gain in buildings. J Sol Energy Eng 131(4):041012
26. Medina MA, King JB, Zhang M (2008) On the heat transfer rate reduction of structural insulated panels (SIPs) outfitted with phase change materials (PCMs). Energy 33(4):667–678
27. Pasupathy A, Athanasius L, Velraj R, Seeniraj R (2008) Experimental investigation and numerical simulation analysis on the thermal performance of a building roof incorporating phase change material (PCM) for thermal management. Appl Therm Eng 28(5–6):556–565
28. Kuznik F, Virgone J (2009) Experimental investigation of wallboard containing phase change material: data for validation of numerical modeling. Energy Build 41(5):561–570
29. Torlak M, Delalić N, Duraković B, Gavranović H (2014) CFD-based assessment of thermal energy storage in phase-change materials—(PCM). In: International energy technologies conference proceedings—ENTECH'2014. Istanbul, Turkey
30. Zukowski M (2007) Experimental study of short term thermal energy storage unit based on enclosed phase change material in polyethylene film bag. Energy Convers Manag 48:166–173
31. Lacroix M (1993) Numerical simulation of a shell-and-tube latent heat thermal energy storage unit. Solar Energy 50:357–367
32. Zhang Y, Faghri A (1996) Semi-analytical solution of thermal energy storage system with conjugate laminar forced convection. Int J Heat Mass Transfer 39(4):717–724
33. Chen C et al (2008) A new kind of phase change material (PCM) for energy-storing wallboard. Energy Build 40(5):882–890
34. Izquierdo-Barrientos M et al (2012) A numerical study of external building walls containing phase change materials (PCM). Appl Thermal Eng 47:73–85
35. Alam M et al (2014) Energy saving potential of phase change materials in major Australian cities. Energy Build 78:192–201
36. Sun X et al (2014) A study on the use of phase change materials (PCMs) in combination with a natural cold source for space cooling in telecommunications base stations (TBSs) in China. Appl Energy 117:95–103
37. Hasan MI et al (2018) Experimental investigation of phase change materials for insulation of residential buildings. Sustain Cities Soc 36:42–58
38. Ahangari M, Maerefat M (2018) An innovative PCM system for thermal comfort improvement and energy demand reduction in building under different climate conditions. Sustain Cities Soc 44:120–129
39. Solgi E et al (2016) Cooling load reduction in office buildings of hot-arid climate, combining phase change materials and night purge ventilation. Renew Energy 85:725–731
40. Álvarez S et al (2013) Building integration of PCM for natural cooling of buildings. Appl Energy 109:514–522
41. Zhou G et al (2011) Energy performance of a hybrid space-cooling system in an office building using SSPCM thermal storage and night ventilation. Sol Energy 85(3):477–485

42. Nagano K et al (2006) Study of a floor supply air conditioning system using granular phase change material to augment building mass thermal storage—heat response in small scale experiments. Energy Build 38(5):436–446
43. Alawadhi E (2008) Thermal analysis of a building brick containing phase change material. Energy Build 40(3):351–357
44. Ning M, Jingyu H, Dongmei P, Shengchun L, Mengjie S (2017) Investigations on thermal environment in residential buildings with PCM embedded in external wall. Energy Procedia 142:1888–1895
45. Faraji M (2017) Numerical study of the thermal behavior of a novel composite PCM/concrete wall. Energy Procedia 139:105–110
46. Solgi E, Hamedani Z, Fernando R, Kari BM, Skates H (2019) A parametric study of phase change material behaviour when used with night ventilation in different climatic zones. Build Environ 147:327–336
47. Navarro L, Solé A, Martín M, Barreneche C, Olivieri L, Tenorio JA, Cabeza LF (2019) Benchmarking of useful phase change materials for a building application. Energy Build 182:45–50
48. Derradji L, Errebai FB, Amara M (2017) Effect of PCM in improving the thermal comfort in buildings. Energy Procedia 107:157–161
49. Panayiotou G, Kalogirou S, Tassou S (2016) Evaluation of the application of phase change materials (PCM) on the envelope of a typical dwelling in the Mediterranean region. Renew Energy 97:24–32
50. Durakovic B, Yildiz G, Yahia ME (2020) Comparatıve performance evaluatıon of conventıonal and renewable thermal ınsulatıon materıals used ın buıldıng envelops. Tehnicki vjesnik—Technical Gazette 27(1) (in press)
51. Hichem N, Noureddine S, Nadia S, Djamila D (2013) Experimental and numerical study of a usual brick filled with PCM to improve the thermal inertia of buildings. Energy Procedia 36:766–775
52. Zastawna-Rumin A, Nowak K (2016) Experimental research of a partition composed of two layers of different types of PCM. Energy Procedia 91:259–268
53. Hasse C, Grenet M, Bontemps A, Dendievel R, Sallée H (2011) Realization, test and modelling of honeycomb wallboards containing a Phase Change material. Energy Build 43(1):232–238
54. Thantong P, Chantawong P (2017) Experimental study of a solar wall collector with PCM towards the natural ventilation of model house. Energy Procedia 138:32–37
55. Kissock J, Kelly J, Hannig M, Thomas I (1998) Testing and simulation of phase change wallboard for thermal storage in buildings. In: Proceedings of international solar energy conference. Albuquerque, New Mexico, 14–17 June 1998
56. Chhugani B, Klinker F, Weinlaeder H, Reim M (2017) Energetic performance of two different PCM wallboards and their assessing the feasibility of using the heat demand-outdoor regeneration behavior in office rooms regeneration behavior in office rooms. Energy Procedia 122:625–630
57. Li Y, Darkwa J, Kokogiannakis G (2017) Heat transfer analysis of an integrated double skin Façade and phase change material blind system. Build Environ 125:111–121
58. Cheng C, Xu L, Liu W, Pingfang H (2018) Thermal simulation of a DOAS + CRCP air conditioning system coupled with a floor containing a phase change material (PCM). IOP Conf Ser Mat Sci Eng 382:1–5
59. Kaltenbach F (2005) PCM latent thermal storage media heating and cooling without energy consumption. Detail Mag 5:544–549
60. Yahaya NA, Ahmad H (2011) Numerical investigation of indoor air temperature with the application of PCM gypsum board as ceiling panels in buildings. Procedia Eng 20:238–248

61. Tokuç A (2015) Experimental data showing the thermal behavior of a flat roof with phase change material. Data Brief V:476–480
62. Hanchi N, Hamza H, Lahjomri J, Oubarra A (2017) Thermal behavior in dynamic regime of a multilayer roof provided with two phase change materials in the case of a local conditioned. Energy Procedia 139:92–97

Chapter 5
PCM-Based Glazing Systems and Components

5.1 Introduction

Glazed units are usually the weak heat barriers which separate the exterior and interior. In cold climates, window glazings are usually responsible for 10–25% of the heat loss from heated ambient. Contrary, in hot climates the excess of solar radiation penetrating through the glass windows usually leads to excessive cooling load, which consumes more electricity [1]. For these reasons, many studies and engineering developments were devoted to creating efficient and thermally effective windows. Especially in recent decades, window technology achieved a relatively high technological standard. Investments in research and development led to a new generation of materials and design options which offer better thermal efficiencies and high performance [2]. There are different advanced systems that allow better control and reduce the heat gain or loss, depending on the design options.

The fact is that window glazing contributes to a substantial part of the heat losses and gains in buildings. Based on the recommendations given in Annex 44 of IEA ECBCS [3], a further investigation on the possibilities of reducing the energy demand related to glazed and/or translucent parts of the facades is necessary.

Traditionally, window sun shading devices were designed as exterior or interior and recently integrated between glazing panes [3]. Even if exterior sun shading systems are preferable, interior systems are installed in many buildings. The reasons range from the sensitivity of exterior systems to surrounding conditions like strong winds, architectural considerations, or simply costs. Conventional interior sun protection systems consist of horizontal or vertical slats, usually with high reflective properties on the side facing the exterior [4]. Improving the thermal mass of these sun shading devices by adding the PCM is possible to manage excessive building heat gain/loss. New scientific discoveries are paving the path to applications of PCM in glazing units instead of outside of them.

© Springer Nature Switzerland AG 2020

B. Duraković, *PCM-Based Building Envelope Systems*, Green Energy and Technology, https://doi.org/10.1007/978-3-030-38335-0_5

5.1.1 Interior Solar-Shading Devices

Blinds can prevent solar radiation into the room behind the blind. To achieve this, blinds can be installed inside the building, or outside in front of the window glazing. Internal blinds cause a thermal problem because solar radiation transmitted through the glazing system is absorbed in the surface of internal blinds, they heat up and release the heat into the room. To handle this issue, Mehling [5] presented the application of the PCM in the interior horizontal blinds applying hydrated salt $CaCl_2 \cdot 6H_2O$. This system is suitable to be utilized under the hot summer climate, especially for those areas with significant daytime and nighttime temperature fluctuations. Another interior sun shading device with vertical slats was studied with similar results. The common problem in both cases is heat gain with the delay. Both studies suggest ventilation at night by opening the window.

Within the project "Innovative PCM technology," funded by the German Ministry of Economics (BMWi), the companies WAREMA and the ZAE Bayern have investigated the idea of reducing and delaying the temperature rise of the blinds by integrating PCM. Figure 5.1 shows the prototype of such internal blinds with PCM, which was developed and tested within the project. Significant daytime and nighttime temperature swing and the attenuation effect on the interior air temperature are observed with the help of PCM sun shading system [5, 6].

Referring to Fig. 5.1, temperature measurements in a test room under realistic conditions have shown promising results: The temperature rise of the blinds decreased by about 10 °C and was delayed by approximately 3 h. The air temperature in the room was about 2 °C lower. Further, investigations by numerical simulation showed a decrease in the operative temperature of the room by about 3 °C and a time shift of the heat release from noon to evening. The thermal comfort during working hours is therefore significantly improved. At night, the heat has to be released to the outside by ventilation.

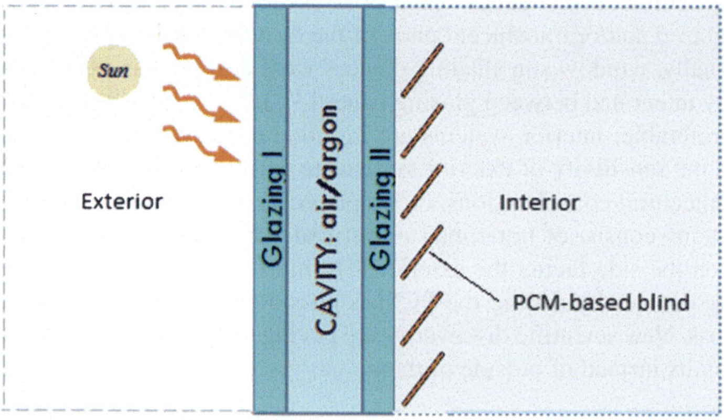

Fig. 5.1 Interior horizontal blind or vertical blind with PCM

Weinlaeder et al. [4] studied another PCM interior sun protection system consisting of vertical slats. Two installed systems, one in westward office rooms located in Karlsruhe and another in southeast office rooms located in Kassel, were monitored from winter 2008 until summer 2010. The authors claim that the system showed a significant cooling potential in summer and even some advantages in winter. In summer mode, the slat surface temperature of the conventional system can often excide 40 °C. The surface temperature of PCM-filled slats hardly ever excided 28 °C on the interior side. To enhance the discharge of the system during the night, they recommended the use of a ventilation system in combination with tilted windows. The left opaque upper part of the facade consists of a ventilation flap for additional airflow during the night.

The pioneer in a sun protection system with PCM was developed by WAREMA, the leading manufacturer of sun shading systems in Germany. The PCM used was Delta®-Cool 28, a salt hydrate with a melting range between 26 and 30 °C and a melting enthalpy of 188 J/g produced by the company Dörken [7, 8]. Dörken also provided the encapsulation of the PCM, a hollow polycarbonate blind with a 12 mm gap filled with the PCM and sealed at the ends, so each square meter of the blind area contained about 17 kg PCM. To ensure a good sun protection performance, the PCM-filled slats were covered with a highly reflective white fabric.

5.1.2 Exterior Solar-Shading Devices

An exterior shutter system containing PCMs is a movable shading element associated with the window facade. In winter mode, the system is to be opened during the day to maximize the solar heat gains indoors through the window glazing. During the night, the system is to be closed to minimize the heat losses through the glazing and vice versa for summer mode. The operation of the system enables the melting of the PCM mass during the day (charging) and its solidification during the night (discharging) by releasing the thermal energy indoors.

Alawadhi [9] in his study used a finite element model for the numerical simulations of thermal analysis of a window with a shutter containing phase change material. Figure 5.2 represents the geometry configuration of windows with a shutter containing PCM. The geometry consists of glass, air gap, and PCM. The outdoor surface of the shutter is subjected to time-dependent solar radiation and forced convection boundary conditions, while the indoor surface of the glass is subjected to time-independent free convection boundary condition.

The thickness of the shutter, L_{sh}, is varied to assess the effect of PCM quantity on the shutter's thermal characteristics, while the thickness of the air gap and glass, L_a and L_g, is maintained constant. Measured radiation data was used at the outdoor surface of the shutter. The thermal effectiveness of the proposed PCM shutter is evaluated by comparing the heat gain at the indoor space to the heat gain of foam

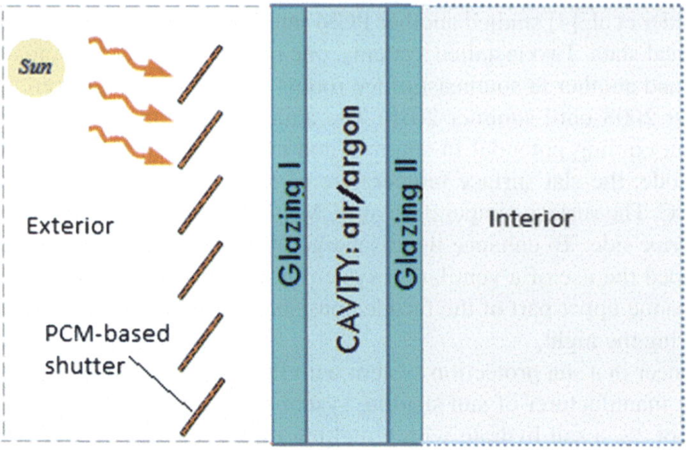

Fig. 5.2 PCM-based exterior shading window shutter

(conventional) shutter during typical working hours. The results show that the magnitude of PCM melting temperature and its quantity in the shutter have a significant effect on the thermal performance of the PCM shutter.

Soares et al. [10] proposed a southward PCM shutters system for winter mode. They used a two-dimensional numerical simulation model. A latent heat storage system has been numerically designed and parametrically optimized to take advantage of solar thermal energy for buildings space heating during the winter in Coimbra, Portugal. An experimental study using full-scale window shutter with PCM also shows the potential for the thermal regulation of indoor spaces [11].

Exterior window shading elements with PCMs were also studied by Buddhi et al. [12]. They studied the thermal performance of a wooden box having a PCM window in the south direction. Commercial grade lauric acid (melting point 42.2 °C, latent heat of fusion 181 kJ/kg) was used as a latent heat storage material. Experiments were conducted during the month of May, and the experimental results were compared with a reference cell. From the results, it was noticed that the PCM window increased the temperature of the test cell during evening and night.

5.1.3 Integral Solar-Shading Devices and Translucent PCM Walls

Ismail and Henriquez studied the possibility of using a window with a movable PCM *curtain* [13]. They performed numerical and experimental study for a thermally effective window using the PCM curtain, as shown in Fig. 5.3. The window was double-glazed with a gap between glazings and an air vent at the top corner. The sides and bottom are sealed with the exception of two holes at the bottom, which are

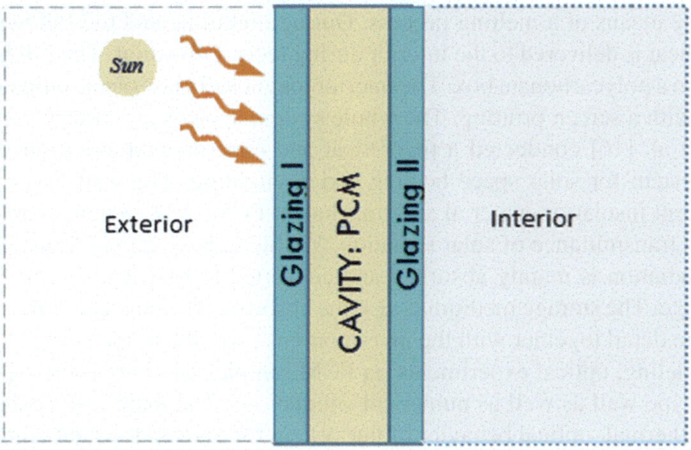

Fig. 5.3 PCM-based glazing system

connected by a plastic tube to a pump and the PCM tank. The pump is connected in turn to the tank containing the PCM, which is in the liquid phase. The pump operation is controlled by a temperature sensor. When the temperature difference reaches a preset value, the pump is operated and the liquid PCM is pumped out of the tank to fill the gap between the glass panes. Because of the lower temperature at the outer surface, the PCM starts to solidify, forming a solid layer that increases in thickness with time and hence prevents the temperature of the internal ambient from decreasing. This process continues until the PCM changes to solid. The authors claim that a *well-designed window system* will ensure that the external temperature will start to increase before the complete solidification of the enclosed PCM.

Authors claim that the external temperature will start to increase before the process of solidification is completed but they did not provide any information in terms of what is "*well-designed window.*" Due to many components included in the window system, it is expensive and expensive for maintenance, complicated to operate for end user and unreliable.

Ismail et al. [1] performed an analysis of two-pane windows with a gas-filled cavity and another with PCM-filled cavity. The double glass window filled with PCM is more thermally effective than the same window filled with air. Similarly, Merker et al. [14] have published more information on shading system development. They have developed a new PCM-shading system to avoid overheating around the window area.

GlassX [15] developed a transparent triple insulating glazing unit providing the system with thermal insulation of a U-value, as they claim, of $U = 0.48$ W/m^{2}K. Light of the low winter sun passes the prism almost unimpeded. Above a certain angle of incidence in the temperate zone between April and September passes the geometry of the prism causing total reflection of the direct sunlight. Then, only the diffuse, low-energy part of the radiation lights the interior. Solar heat is stored in the PCM by means of a melting process. During nighttime and the following days, the stored heat is delivered to the interior during recrystallization. The salt hydrate is contained in a polycarbonate box. The interior toughened glass pane could optionally be coated with a screen printing. The whole system appears as a translucent wall.

Manz et al. [16] conducted a theoretical and experimental study on the external wall system for solar space heating and daylighting. The wall was composed of transparent insulation material and translucent PCM. This system enables selective optical transmittance of solar radiation. Visible light is mainly transmitted, and invisible radiation is mainly absorbed and converted to heat, causing in particular phase change. The storage medium is also the absorber. The concept of the system is presented in detail together with the investigations carried out, including a brief outline of modeling, optical experiments on PCM samples and long-term experiments on a prototype wall as well as numerical simulations. The main results showed the promising thermal–optical behavior of the system for a Swiss lowland climate, even during the month with the lowest irradiation. The parameters of the prototype wall with a mean melting temperature of the PCM of 26.5 °C were assumed. When considering the percentage of time in which the building does not lose energy through the south-facing wall, a maximum can be reached with a mean melting temperature of approximately 20–21 °C. In this case, energy losses through the facade occur only during 1% of the time.

Bontemps et al. [17] conducted an experimental and numerical simulation study of the application of phase change materials in building a wall made of hollow glass bricks filled with PCM. The test was carried out in real climatic conditions in an outdoor test cell constituted of two small rooms separated with a wall containing PCM. A specific wall made of hollow glass bricks filled with PCM was studied. Three PCMs were tested: fatty acid, paraffin, and salt hydrate whose melting temperatures are 21 °C, 25 °C, and 27.5 °C, respectively. Indoor and outdoor temperatures were measured with thermocouples, and flux meters were located at the center of each wall as well. Reasonable agreement between the simulation and the experimental results was observed, and the authors pointed out the importance of conceiving systems with PCMs coupled with efficient night ventilation.

DELTA®-COOL 28 is applied in translucent polymethyl methacrylate (PMMA) panels in a glass facade system of a zero-energy office building in Kempen, Switzerland. Every second window panel is equipped with the phase change material in order to reduce solar heating of the interior office space. The solar energy that is stored in the PCM gets release during nighttime when typically the building would need to be heated to maintain a desirable temperature [18].

Goia et al. [19–21] performed an experimental analysis on a double-glazing system with paraffin wax. They performed an outdoor test in a cell facility located in a temperate subcontinental climate. Also, they developed a simplified numerical model of PCM glazing system, and simulations were compared against experimental data on a simple PCM glazing. They concluded that the numerical tool seems to predict pretty well the overall thermal behavior and can be used to simulate PCM glazing systems [20]. The surface temperatures and the transmitted irradiances of the PCM glazing prototype and of a reference fenestration, measured over a six-month experimental campaign, have been used to numerically evaluate the indoor thermal conditions inside a typical office room. Different boundary conditions, ranging from summer to winter season, including the mid-season, have been analyzed [21].

Weinläder et al. [22] studied a PCM-facade-panel for daylighting and room heating. The double-glazing facade combined with PCMs as a third layer is able to transmit enough light and have a more equalized energy balance during the day in comparison with a double-glazing facade without PCM. The results showed that this system could be a good choice for lightweight construction. In winter, especially during evenings, the PCM-facade-panel provided homogeneous illumination and fewer heat losses improving the thermal comfort. In summer, the results showed a low heat gain, which reduces peak cooling loads during the day.

Gowreesunker et al. [23] studied the optical and thermal properties of a small-scale RT27 PCM-glazed unit. They studied optical and thermal performance by a combined experimental–numerical CFD model analysis.

The impact of optical properties on the thermal performance of the PCM-filled double-glazing system is notable, and the effect of the PCM phase is also strong [24]. Different thermo-physical parameters of PCM such as latent heat of fusion can enhance the thermal energy storage capacity of the glazing system [25]. Spectral and angular solar properties of integrated double-glazed system in the range of 400–2000 nm show highly scattering effect in the solid phase of the PCM layer, with increasing weight of the direct-to-scattering transmission mode as the PCM layer thickness increases [26], while radiation absorption dominates in the liquid phase [23].

Double-glazed PCM system to be suitable for summer and winter mode interchangeably must use PCMs with phase transition temperature lower than the room temperature in winter [27]. Otherwise, experimental and numerical studies show that thermal insulation and load shifting effects will reduce energy consumption only for typical sunny and rainy summer days [28, 29]. Relatively long periods of sun combined with high exterior temperatures are required for phase transition in full-scale PCM-filled window. Thus, the low potential of application of the window system in the Nordic climate is presented [30, 31]. The thermal performance of the glazed system can be increased by increasing the latent heat of fusion of a PCM and by selecting temperature transition from the range of 25–31 °C [32].

Duraković and Torlak [3] investigated the impact of the cavity thickness on the glazing units using both experimental [33] and numerical methods on a double-glazed window. In the natural setup, the exposure was done naturally having both chambers represent the interior. In the laboratory environment, Chamber #1 represents the exterior while Chamber #2 interior [34]. In the computational model represented by the model, loses or gains heat from both ambient by conduction, convection, and radiation [3]. The glass thickness is 4 mm, and the cavity between the glass plates is between 6 and 30 mm. The weather conditions are taken into consideration typical values holding on September 22 for Sarajevo, Bosnia, and Herzegovina [3]. Both computed and measured temperatures seem to be approximately the same giving legitimacy to the findings of the paper. The decrease in temperature can be accommodated by a proper cavity thickness. It was found that a 24-mm-cavity size is optimum and determined by liquid volume fraction, which cannot solidify overnight. Therefore, it is meaningless to increase cavity size over 24 mm since any amount beyond that portion will not get a chance to solidify overnight [3]. No more than 19 mm is recommended in glazing limited to the radiation, meaning the north-facing glazing [3]. The results of this study are also compared with the previous studies and discussed in Table 5.1.

Table 5.1 Previous study results and comparison [35]

	Author/reference	Phase transition temperature period (h)	Transition/peak surface temperature (°C)	Comment
Interior	ZAE Bayern [6]	5	27/28	Relatively shorter transition period of 5 h at 27 °C. Since the blind is placed in the room, whole heat stays trapped in interior—no cooling outside, (causes late heat gain)
	Weinlaeder et al. [4]	7	28/32	Relatively shorter transition period of 7 h at 28 °C. Since the blind is placed in the room, whole heat stays trapped in interior—no cooling outside, (causes late heat gain)
Integrated	Zhong et al. [29, 32]	6	31/32	Transition temp. is 31 °C and thus requires cooling during the transition period of 6 h. Partial cooling outside is enabled
	Ismail et al. [1]	5	24/30	Shorter transition period of 5 h. Partial cooling outside is enabled
	Li et al. [24, 25]	10	28/32	10 h of transition period around 28 °C. Partial cooling outside is enabled
	Optimum PCM thickness [35]	17	26/32	Provides the longest transition period of 17 h at surf. temp. of 26 °C. Partial cooling outside is enabled
Exterior	Silva et al. [11]	4	21/37	4 h of transition period around 26 °C

This table was published in Duraković and Torlak [3]

5.2 PCM-Based Versus Conventional Glazing System

Variation of the interior glass surface temperature is notable in preventing building heat gain/loss. Therefore, the temperature variance in conducted experiments and simulations is monitored and recorded with the aim of estimating its impact on the heat gain/loss. The results are discussed separately in the following sections.

The following series of experiments have been done in natural environment. Exterior variables such as air temperature and irradiance have been recorded in the state as they had appeared. The noise presented in some period is mostly caused by the presence of clouds, which reflects on the interior surface temperature history (Fig. 5.4). In most of the cases, the clouds were present in the afternoon when solar irradiance had decreasing trajectory.

Referring to Fig. 5.4, two samples with different materials in the cavity (air and PCM) had been exposed to the environment simultaneously. The temperature changes on the interior glass surfaces ($T_{4\text{air-e}}$ and $T_{4\text{pcm-e}}$) caused by irradiance ($I_{\text{n-e}}$) and exterior air temperature ($T_{\text{ex-e}}$) from the environment had been recorded. Air sample temperature on the interior surface ($T_{4\text{air-e}}$) is more sensible to the change of the sun radiation flux. Compared to the PCM sample, the air sample (conventional glazing) maintains higher temperature for a few degrees in the period the PCM *melting process start* to the *solidification process start*. This period is considered as heat gain period, which means that the conventional glazing (with air-filled cavity) admits more solar

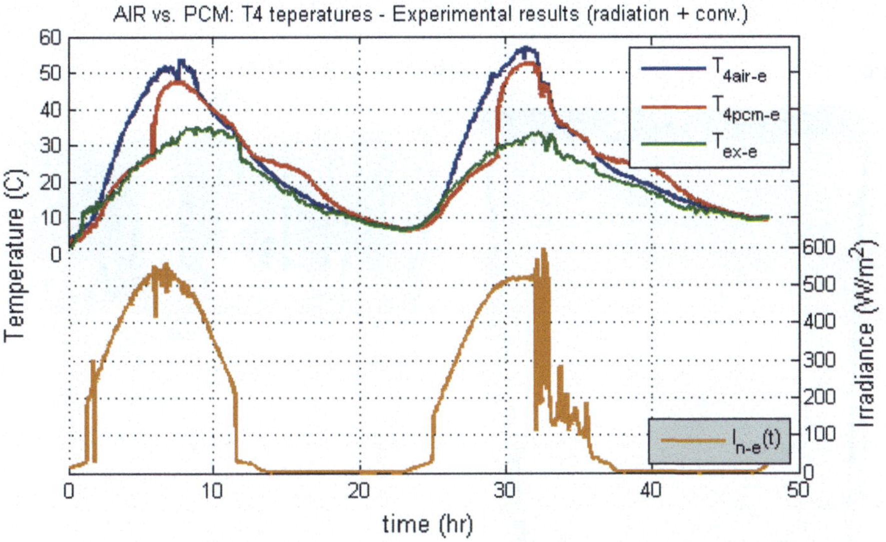

Fig. 5.4 Experimental results of interior surface glazing temperatures caused by solar radiation and convection in natural setting ($T_{4\text{air}}$—air-filled cavity; $T_{4\text{pcm}}$—PCM-filled cavity; T_{ex}—exterior air temp.; I_{n}—irradiance on the exterior glass surface; subscript "e" denotes experimental results), (This figure was published in Duraković and Torlak [3])

irradiance to the interior than the PCM-filled glazing. Overall, the peak temperature associated with the air sample is higher for cc 3 °C, and solar radiation affects the interior as soon as it appears on the surface of the glass immediately with no temperature shift. All of the above qualifies glazings in the traditional design as lower performance products compared to the glazing using new approaches in design by selecting the right materials and implementing them (e.g., phase change materials). In the case of the PCM sample, there are approximately three hours of temperature delay during the melting process caused by latent heat that absorbs the heat in the melting process. This melting process protects the interior from overheating for a given period. At the end of the melting process, sun rays through the liquid phase reached surface temperature sensor easily and instantaneous temperature jump is observed. The PCM was able to absorb the heat and protect the interior glass surface from overheating until the last grain of the solid phase was presented. Once the solid phase is melted completely, the surface temperature was instantaneously raised for couple of degrees and overheating of the interior glass surface and the interior started. The peak temperature of the air sample compared to the PCM sample was about three degrees (Fig. 5.4). When the exterior air temperature is above phase change temperature, the temperature-driven heat, as well as solar radiation, is absorbed by the PCM keeping the interior glass surface temperature in the vicinity of phase transition temperature for the time needed to melt the PCM. Once the PCM is completely melted, there is the instantaneous jump of heat gain mainly due to radiation.

Visualization of phase transition process between glazings caused by solar radiation during the experimental setup is shown in Fig. 5.5.

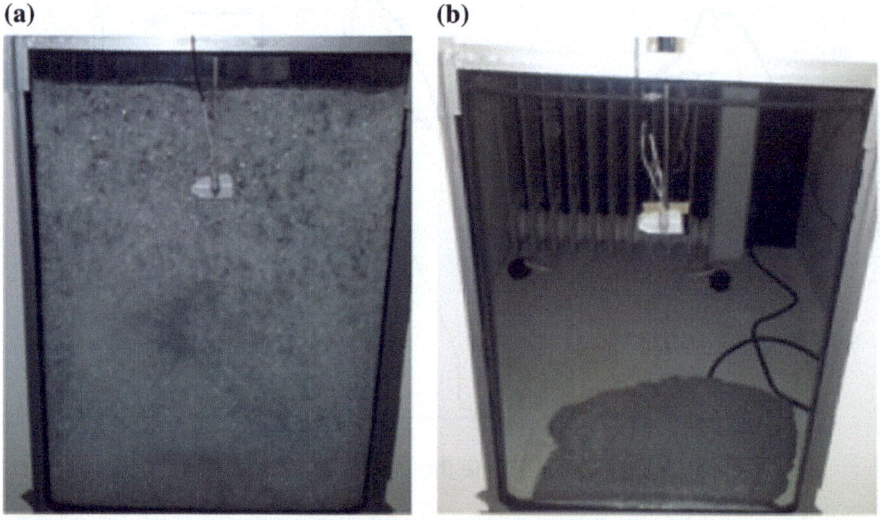

Fig. 5.5 PCM melting process visualization: **a** experiment start—translucent glazing; **b** melting process is completed—transparent glazing (Original author's photo)

Figure 5.5a was taken right after the experimental setup and shows the start of the phase transition where the small portion of the liquid phase is observed in the cavity. Figure 5.5b shows the solid and liquid phases where the solid phase stays on the bottom during the process of melting. It is caused by differences between densities of the solid phase and liquid phase (solid density = 880 kg/m^3; liquid density = 750 kg/m^3). This fact tells us that there is no uniform heat transfer through the glazing area and the temperature distribution over the solid and liquid phases on the glazing area varies. The liquid phase is characterized by increased thermal conductivity compared to air (about 10 times) and therefore is noticed and sudden temperature rise ($T_{4\text{pcm-e}}$) shown in Fig. 5.4.

5.3 Computational Results for Each Glazed Surface

Regarding the previously mentioned, a numerical simulation model has been defined for the prediction of the interior glazing surface temperature, the exterior temperature, and irradiance for all three materials. Figure 5.6 displays experimental versus simulation results with the aim of the validation numerical model, which is used further for the simulation of various cavity widths.

Referring to Fig. 5.6, daily fluctuation of the solar irradiance ($I_{\text{n-s}}$) and exterior air temperature ($T_{\text{ex-s}}$) has been mathematically described and used for exterior boundary conditions. The fluctuation of the solar radiation with its maximum daily value of 530 W/m^2 on the vertical plane is well described by Eq. (5.1) and graphically displayed in Fig. 5.6.

$$I_{\text{n-s}}(t) = I_{\text{n-s,max}} \sin\left(\frac{t}{3600 \cdot \omega} \cdot 2\pi\right) + D \tag{5.1}$$

where $I_{\text{n-s}}(t)$ is the incident solar radiation at any given time on the vertical plane (exterior glass surface) in W/m^2, $I_{\text{n-s,max}}$ is the maximum solar radiation per a cycle W/m^2, t is the time in s, ω is a day period in hours, D is the vertical shift of the function ($D = 80$ W/m^2) to obtain the appropriate daylight hours typical for Sarajevo at the beginning of June. Irradiance simulation has been carried out for the following conditions:

$$I_{\text{n-s}}(t)\begin{cases} > 0, => I_{\text{n-s}} = I_{\text{n-s}}(t), \text{dayligt hours} \\ \leq 0, => I_{\text{n-s}} = 0, \text{at night} \end{cases} \tag{5.2}$$

If the noise caused by the clouds is removed from the recorded irradiance on the exterior vertical glazing surface ($I_{\text{n-e}}(t)$), a good matching between the recorded and the calculated function described by Eqs. (5.1) and (5.2) is noticed.

Referring to Fig. 5.6, exterior air temperature history ($T_{\text{ex-e}}$) is recorded and approximately described by the exterior air temperature function ($T_{\text{ex-s}}$) Eq. (5.3) and further used in the simulation. This ($T_{\text{ex-e}}$) temperature causes convective component

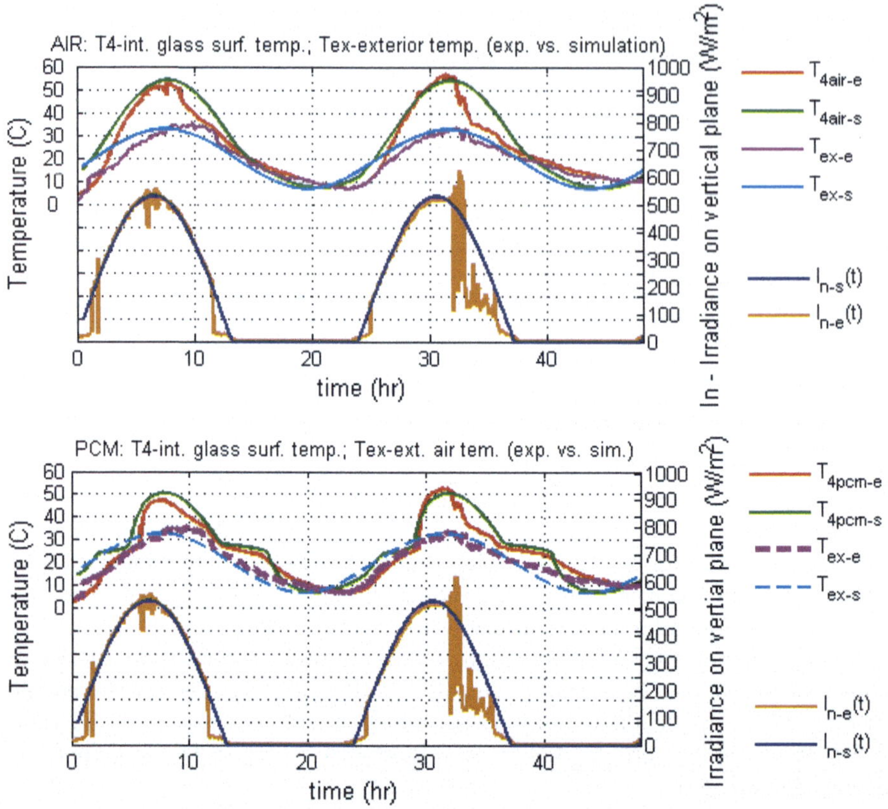

Fig. 5.6 Verification of numerical model with experimental results: T_4—temperatures ($T_{4\text{air-e}}$—air-filled cavity experimental results; $T_{4\text{air-s}}$—air-filled cavity simulation results; $T_{4\text{H}_2\text{O-e}}$—water-filled cavity experimental results; $T_{4\text{H}_2\text{O-s}}$—water-filled cavity simulation results; $T_{4\text{pcm-e}}$—PCM-filled cavity experimental results; $T_{4\text{pcm-s}}$—PCM-filled cavity simulation results; $T_{\text{ex-e}}$—recorded exterior air temp. experimental; $T_{\text{ex-e}}$—exterior air temp. simulation; $I_{\text{n-e}}(t)$—irradiance on the exterior glass surface experimentally recorded; $I_{\text{n-s}}(t)$—irradiance on the exterior glass surface by simulation (verification on 12 mm cavity)), (This figure was published in Duraković and Torlak [3])

of the heat transfer which is slightly small [36]; thus, it is considered as acceptable matching between these two curves for further analyses.

$$T_{\text{ex-s}}(t) = \frac{1}{2}\left(T_{\text{ex},\max} - T_{\text{ex},\min}\right) \cdot \sin\left(\frac{t}{3600 \cdot \omega} \cdot 2\pi\right) + \frac{1}{2}\left(T_{\text{ex},\max} + T_{\text{ex},\min}\right) \tag{5.3}$$

where $T_{\text{ex-}s}(t)$ is calculated value of the exterior air temperature at any given time in °C, $T_{\text{ex,max}}$ and $T_{\text{ex,min}}$ are maximum and minimum exterior air temperature uses in simulation in °C, ω is a day period in hours, $(T_{\text{ex,max}} - T_{\text{ex,min}})/2$ is temperature amplitude, $(T_{\text{ex,max}} + T_{\text{ex,min}})/2$ is vertical shift of the $T_{\text{ex-s}}(t)$ function.

Including in the simulation model previously defined solar radiation and exterior air temperature change defined by Eqs. (5.1), (5.2), and (5.3), and referring to Fig. 5.6, the simulation temperature response on the interior glass surface for different materials in the cavity ($T_{4air\text{-}s}$, $T_{4pcm\text{-}s}$) provides a similar trend as the recorded temperature history during the experiment ($T_{4air\text{-}e}$, $T_{4pcm\text{-}e}$). Recorded temperatures have some deviations from day to day caused by clouds and the exterior air temperature fluctuation ($T_{ex\text{-}e}$) but the overall computed and measured temperature history responses on the interior glass surface have an acceptable matching trend. Thus, the computational model can be used as a good predictor of the glazing system performances for different cavity sizes.

In further computations, regional and weather parameters have been readjusted to fit average conditions on a typical summer day. The following parameters have been used for the computation of the various sizes of the system performances: $T_{e,max} = 34$ °C; $T_{e,min} = 17$ °C; $h_{ex} = 25$ W/m^{2}K; and cavity widths from 6 mm to 30 mm. The temperature histories have been simulated for external glass surface, midplane of the cavity, internal glass surface, with a convective boundary condition at the internal glass surface and presence of solar radiation along with convection on the external glass surface. A typical representative of the results obtained by numerical simulation for the cavity width of 26 mm for all considered materials has been presented in Fig. 5.7.

The cavity size of 26 mm represents a specific case of the solidification process. The heat is removed by convection only during the solidification process and due to small temperature differences between the glass surfaces and its environment, as result, some portion of the liquid phase remained at the end of the solidification process. The solidification process takes a longer time to cool accumulated heat by radiation and convection than it can be available in reality. Glazing system performances for each material particularly are shown in Fig. 5.8.

The water temperature shift is observed, but the peak temperature is slightly higher than the air. The water continues heating the space for a while after the sun sets and the exterior air temperature falls below the interior temperature. This effect is harmful to the interior thermal comfort due to increased temperature differences between the glass surface and interior air. The PCM temperature flattening and the peak temperature on the interior glass surface over the heating cycle are notable, and these effects may contribute to the thermal comfort of the interior space. In the combined effect of solar radiation and convection, the heating phase causes a shorter period of melting than the solidification.

5.4 Material Performances for Various Environments

To analyze different materials as design components in the glazing system, three basic aspects are considered: thermal conductivity of the material, heat gain, and the period that the materials can contribute to relaxing of A/C system.

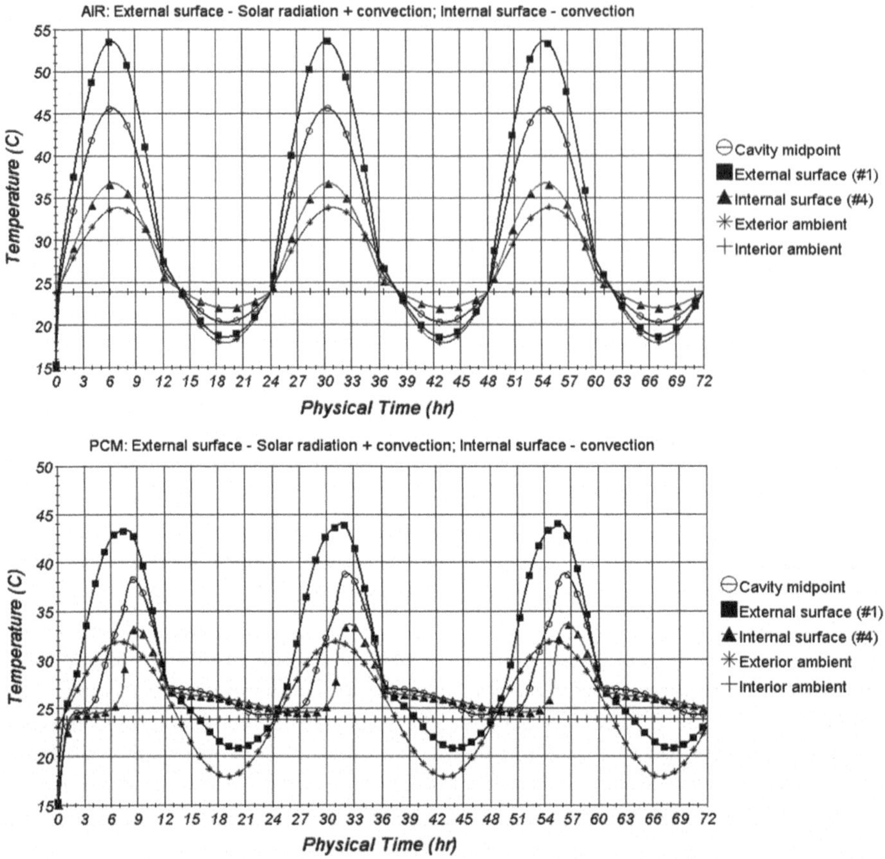

Fig. 5.7 Simulation results: temperature history at three different planes of double-glazed unit filled with air (top), and PCM (bottom), obtained with convective heat flux at the internal glass surface and presence of solar radiation on the external glass surface for cavity width of 26 mm (Original author's photo)

In this study, these three materials have significantly different capabilities for conductive heat transfer and accumulation of energy. As the insulator, the air with its conductivity of 0.02 W/mK is the best heat transfer insulator and then follows the PCM with its conductivity of 0.2 W/mK. The heat transfer coefficients (*U*-values) of the PCM and the air glazing for various cavity widths were calculated per the following equation:

$$U = \frac{1}{\frac{1}{h_e} + \frac{L_g}{\lambda_g} + \frac{L_{cav}}{\lambda_{cav}} + \frac{L_g}{\lambda_g} + \frac{1}{h_i}} \tag{5.4}$$

where L_g is glass thickness in m, L_{cav} is cavity thickness in m, λ_g is glass conductivity W/mK, λ_{cav} is conductivity of the material in the cavity (air or PCM) in W/mK, h_{ex}

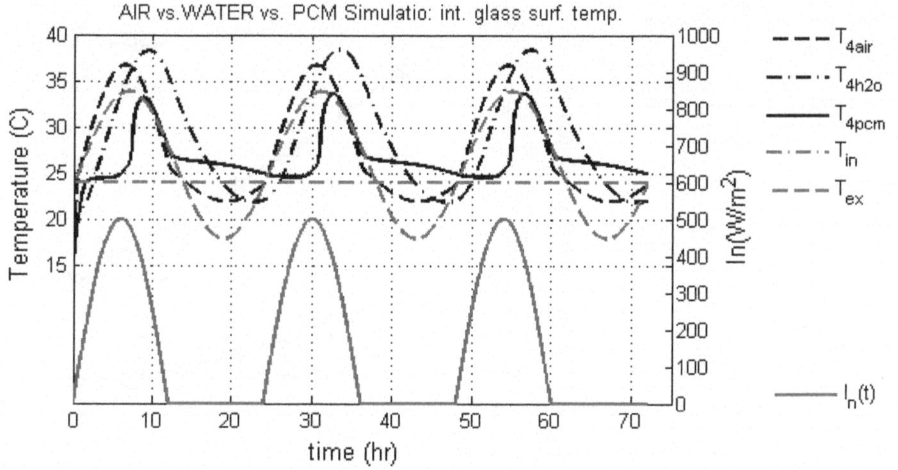

Fig. 5.8 Simulation results: T_4 temperature history of double-glazed unit filled with air $T_{4\text{air}}$, water $T_{4\text{H}_2\text{O}}$, and PCM $T_{4\text{pcm}}$, obtained with convective heat flux at the internal glass surface and presence of solar radiation I_n on the external glass surface for cavity width of 26 mm (Original author's photo)

and h_in are the convective coefficients on the exterior and interior glass surface. The results of the calculation are displayed in Fig. 5.9.

Material conductivity is not the only assessing parameter, specific heat capacity must be considered as well. Referring to Fig. 5.9, it is obvious that PCM has higher U-values but the heat storage capability of each material is notably different. The material stores the heat and later releases to the environment reducing the temperature variability which has a positive impact on the A/C load. The heat gain depends on U-value, temperature, the time period exposed to the heat source, as well as of the stored heat in the material. Total heat energy gain over a day cycle per a square meter area under-considered conditions is calculated as:

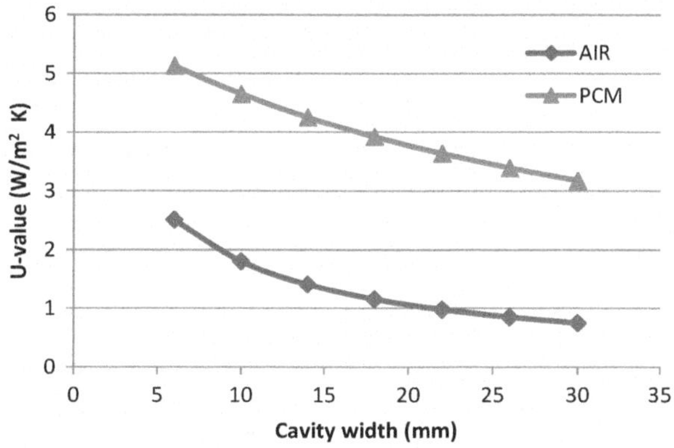

Fig. 5.9 U-value for different materials in the cavity (Original author's photo)

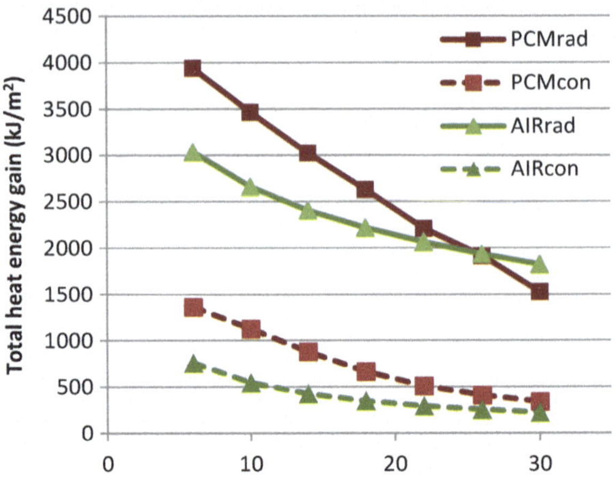

Fig. 5.10 Total heat energy gain caused by convection/radiation and convection (Original author's photo)

$$q = \alpha_{\mathrm{i}} \int_{0}^{24} (T_4(t) - T_{\mathrm{in}})\mathrm{d}t \tag{5.5}$$

The results are shown in Fig. 5.10 for all materials with different cavity widths. The figure displays compared results of the heat energy gain due to solar radiation and convection (solid lines) and the case when the convection is presented only (dashed lines).

It was observed from the figure that the heat gain to the interior is the highest in the case of the PCM, and at the end comes the air for both heat transfer modes. If only judged by the amount of heat energy that enters the space, then it could be said that the air is the better material for both heat transfer modes.

On the other hand, taking into account the time period during which the heat is transferred, it is not possible to make the same conclusion. The length of the heat gain period in a day cycle varies from material to material. If the heat gain period is shorter, it may cause fast overheating otherwise the longer period means less heat per unit of the time. Figure 5.11 shows the heat gain curves per unit of time for the various materials and cavity widths over the heat gain period. The curves are obtained by dividing the total heat energy gain from Fig. 5.10 by the heat gain period. The heat gain period is determined as the period in which interior glass surface temperature (T_4) stays above $T_{\mathrm{in}} = 24$ °C. This period varies depending on the material between glazings. Due to its latent heat, the PCM has the longest period that keeps interior glass surface temperature above interior air temperature. Therefore, heat gain per unit of time through the PCM glazing will be notably reduced compared to the air-filled glazing [37].

In convection mode only (dashed lines), the air provides a lower amount of heat gain. In the presence of solar radiation together with the convection (solid lines), the PCM provides the smallest portion of the heat gain per unit of time for cavity widths over 12 mm and has a trend of decreasing the heat gain with the increase of the cavity. This can be explained by the latent heat of the PCM that provides longer heat

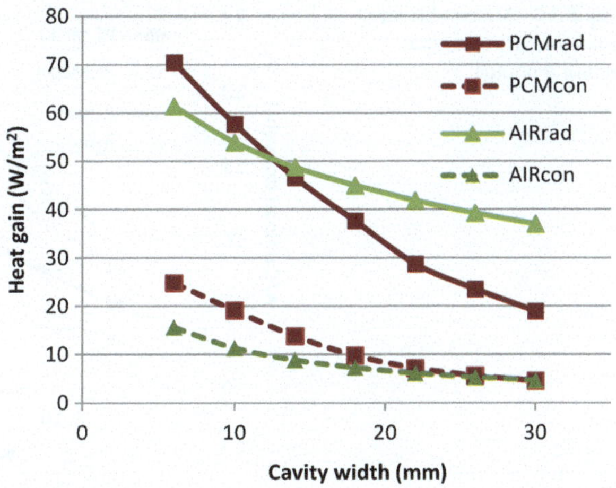

Fig. 5.11 Heat gain per unit of time for various cavity widths caused by convection/radiation and convection (Original author's photo)

gain to the interior reducing the temperature swing. It is obvious that temperature-driven heat transfer mode is not dominant in case of glazing systems. The radiative component is about four times higher than the temperature-driven component for each material and thus must be considered first in glazing system design.

The other aspect of the design is the time period that the materials used in design may contribute to relaxing of air conditioning (A/C) load (called *A/C unload time*). It is assumed that A/C system turns on/off in a temperature interval up to 3 °C, which depends on the A/C device. A/C unload time is defined as the time needed to increase interior glass surface temperature from 23 to 26 °C, while the interior air temperature remains 24 °C. This is the temperature interval that may contribute to the A/C system in reducing the load period. The A/C unload period for each material in the cavity is determined as:

$$t_{\text{unl}} = t(T_4) \text{ for} \begin{cases} T_4 > 23\,^\circ\text{C} \\ T_4 < 26\,^\circ\text{C} \end{cases} \tag{5.6}$$

Figure 5.12 represents hours in case there is a contribution from the glazing system to A/C unload for radiative and temperature-driven heat transfer modes.

In the absence of solar radiation, there is an interaction between the air and the PCM. The air performs better with a smaller cavity width while the PCM performs better with the higher cavity widths. The slope of the air curve changes in the vicinity of a 15 mm cavity, which is caused by air convection between the glazings. Convection fosters the heat transfer between plates causing a reduction in unload time by changing the slope of the curve. The slope of the PCM curve changes in the vicinity of a 16 mm cavity, which is caused by the remaining solid volume fraction for the cavity over 16 mm. The remaining solid volume fraction of the PCM causes a reduction in heat storage trend and consequently reduces unload time. Since the

Fig. 5.12 Air-conditioning unload hours (Original author's photo)

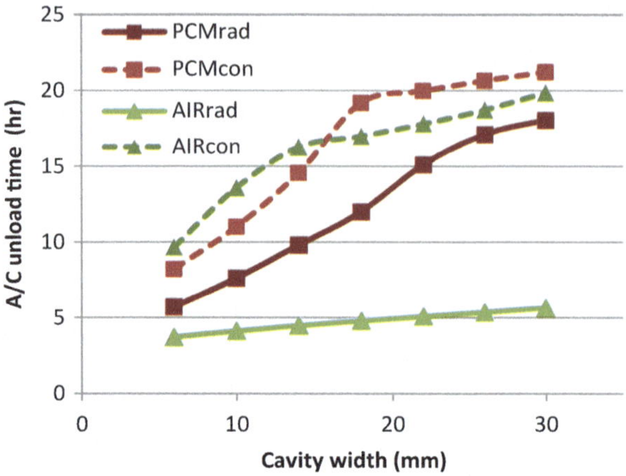

convection mode is rare and not dominant in real life, the sun radiation component plays the main role.

In cases, the radiation component of the heat transfer is presented, and the PCM notably has the longest time period during which the A/C can be off, qualifying the PCM as better material in the glazing system design. This is caused by the higher heat storage capacity of the PCM in the form of the latent heat. The slope of the PCM curve changes in the vicinity of 25 mm causing a reduction in unload time. The slope change is caused by the remaining liquid volume fraction during the solidification process. The slope of the air trajectory is low below the PCM trajectory. Thus, the air is regarded as lower performance material than the PCM in the presence of the radiation, which is a general case in reality.

The recommendations and trends of the curves from the previous diagrams are discussed in detail in the following section.

5.5 Design and Decision Criteria

Here are in detail described design and decision criteria of environmentally responsive glazings based on the PCM, which may contribute to the attenuation of the temperature fluctuation and preventing a building from overheating. Since design and decision criteria depend on the direction glazing as well, two main groups regarding the direction were studied and recommended:

North-facing glazings where the impact of the solar radiation is negligible (convection and conduction heat transfer modes are dominant).

Glazings facing the sunny sides—they are *oriented* on the *sunny side* of the building from the east over the south to the west.

5.5.1 Sunny-Side Orientation Glazed Unit

Glazing orientation affects the building's thermal state by transmitting solar radiation directly into the conditioned interior space. The heat is trapped inside the space and heats the interior surfaces causing the "greenhouse effect," which is beneficial during winter time but undesirable during summer when it can overheat the space, putting more load on the air conditioning system and wasting the energy as well.

It is the summer mode with the peak of the solar radiation on the exterior vertical glass surface of 500 W/m^2, combined with the temperature-driven heat gain. The exterior temperature changes from 17 to 34 °C in a period of 24 h, which represents typical summer day conditions for Bosnia. The indoor temperature is kept constant at 24 °C the whole period. To carry out this analysis, the linear multivariate regression model is applied to describe the relationship between cavity width and the temperature oscillation, which is defined by the following equations:

Average temperature

$$T_{4,\mathrm{ave}} = -0.094x + 28.74 \tag{5.7}$$

Temperature variation

$$T_{4,\mathrm{StDev}} = -0.211x + 8.503 \tag{5.8}$$

Remaining liquid ratio after solidification process

$$f_{\mathrm{L/S}} = 0.026x - 0.651 \tag{5.9}$$

where x is cavity width in mm.

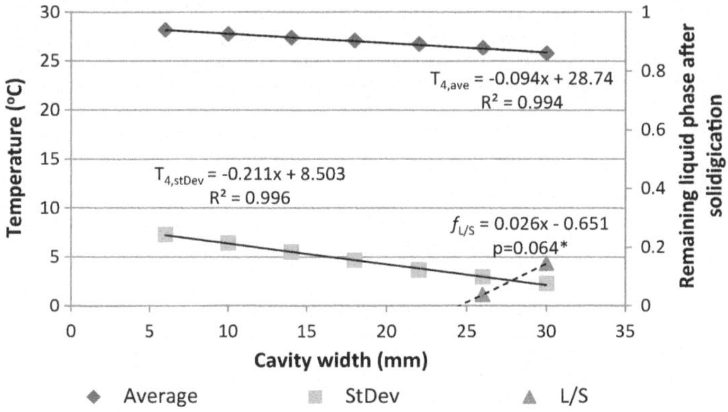

Fig. 5.13 Radiation and temperature-driven heat transfer: standardized temperature deviation as design criterion (Original author's photo)

Figure 5.13 shows significance of the temperature swing, average temperature, melting/solidification ratio for various cavities.

There is a strong negative relationship between the cavity and average temperature as well as temperature variation and the cavity. The functions are approximated with linear functions with high coefficients of determination ($R^2 = 0.994$ and $R^2 = 0.996$, respectively). The melting process is fostered by the sun radiation on the exterior glass surface, which completely turns the solid phase to the liquid phase for each cavity size during the solidification period. On the other hand, the solidification processes were not completed for some cavities. The heat transfer mechanism in the solidification process was convection only which takes longer time to cool accumulated heat by radiation and convection from the outside. Longwave radiation during the solidification phase is neglected due to small temperature differences. Thus, the remaining liquid phase after the solidification period is locally linear after a 25 mm cavity, and the relationship is not significant at the significance level of 5% but it is significant at the significance level of 10%.

Heat gain and the A/C load/unload time are considered as other criteria for the design of the responsive glazing system. The times and the radiation heat gain can be described good enough by linear functions. Figures 5.14 and 5.15 show the relationship between the cavity and the design criteria.

There is a significant positive effect of the cavity size on the unload time and the charging time, which is characterized by a linear relationship with the high value of the coefficient of determination ($R^2 > 0.9$). This is beneficial for the thermal comfort because the phase transition temperature, which is close to the comfort temperature, stays for a longer time on the interior surface of the glass. Also, more charging time the more heat is accumulated for later usage during the solidification phase, which keeps the temperature at the interior glass surface for a longer time as well. On the

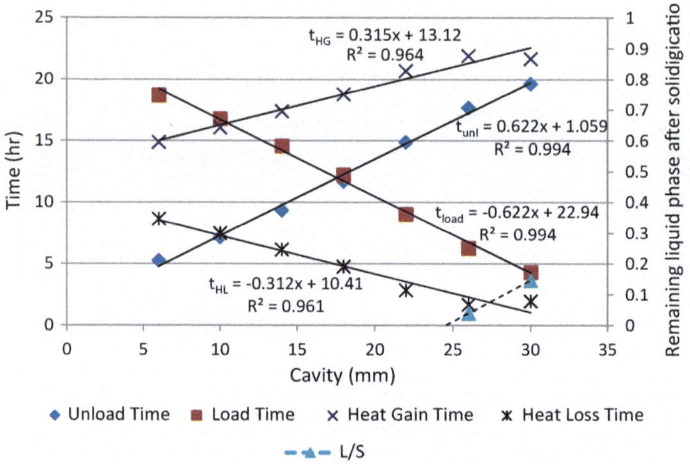

Fig. 5.14 Load versus unload time; charging versus discharging time (Original author's photo)

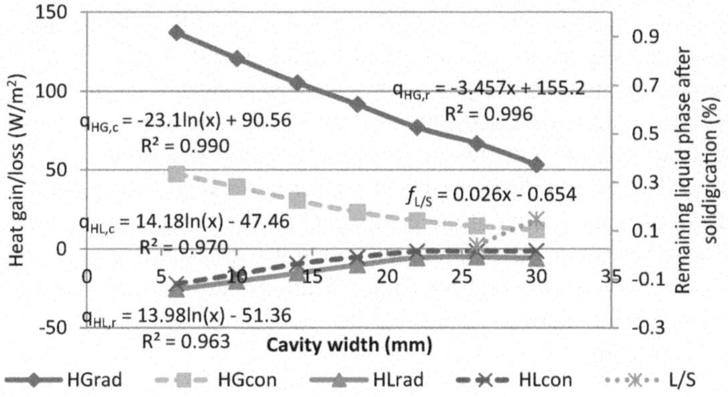

Fig. 5.15 Convective and radiative Heat gain (HG) versus heat loss (HL) (Original author's photo)

other hand, there is a negative effect of the cavity size on the load time and heat loss time with $R^2 > 0.9$. The benefit of this can be utilized for the cavity widths less than 25 mm.

The heat gain time and heat loss time have locally linear relationship for all cavity sizes less than 25 mm, after which the function changes the slope. The slope changing is caused due to remaining of a liquid-phase portion after solidification period. The solidification period ends before the solidification process is completed. This is the point at which designers can identify the cost-effective design. Again, going over 25 mm with the cavity increases the cost and makes the design heavier, and the benefit of the discharging heat will not be utilized completely. These relationships have been described by Eqs. (5.10), (5.11), (5.12), and (5.13).

Unload time

$$t_{\text{unl}} = 0.622x + 1.059 \tag{5.10}$$

Load time

$$t_{\text{load}} = -0.622x + 22.94 \tag{5.11}$$

Heat gain time

$$t_{\text{HG}} = 0.315x + 13.12 \tag{5.12}$$

Heat loss time

$$t_{\text{HL}} = -0.312x + 10.41 \tag{5.13}$$

Heat gain/loss analyses for the PCM glazing exposed to the average summer environmental conditions are an important aspect of the design. The relationship between the size of the cavity and the heat gain/loss is described in Fig. 5.15.

Radiative heat gain is described the best by linear function while the convective components have nonlinear characteristics. Expectation functions of the convective components of the heat gain/loss are the best described with the logarithmic functions and indicate significant effects of the cavity size on the heat gain/loss with high R^2-values. Functions are distinctly curved in the vicinity of 25-mm-cavity widths, which again leads to the conclusion that the cavity width greater than 25 mm significantly differ per this criterion as well.

Radiative heat gain

$$q_{\mathrm{HG,r}} = -3.457x + 155.2 \tag{5.14}$$

Convective heat gain

$$q_{\mathrm{HG,c}} = -23.1\ln(x) + 90.56 \tag{5.15}$$

Heat loss with the presence of the radiative component during the charging period

$$q_{\mathrm{HL,r}} = 13.98\ln(x) - 51.36 \tag{5.16}$$

Heat loss with the absence of the radiative component during the charging period

$$q_{\mathrm{HL,c}} = 14.18\ln(x) - 47.46 \tag{5.17}$$

As mentioned earlier, the reason lies in the rest of the liquid phase that has failed to complete the process of the solidification, which takes place during the convective heat transfer mechanism. This happens due to the shortness of the solidification period required amount of the PCM contained in the cavity over 25 mm and leads to the conclusion that a 25 mm cavity is recommended for the design.

5.5.2 North-Facing Glazed Unit

As design criteria and decision criteria here are studied the following: glazing surface temperature, the time that has no contribution to A/C load, solid/liquid fraction of the PCM. For the north-facing glazings, in this case, temperature-driven heat transfer is counted only. Diffuse radiation is neglected due to the low contribution to the heat gain. The temperature swing for the various cavity widths is shown in Fig. 5.16.

Outside temperature changes per sine function while the indoor temperature is kept constant for 24 °C. Melting and solidification curves represent the temperature on the interior glass surface for various cavity widths. The curves are fitted for each cavity

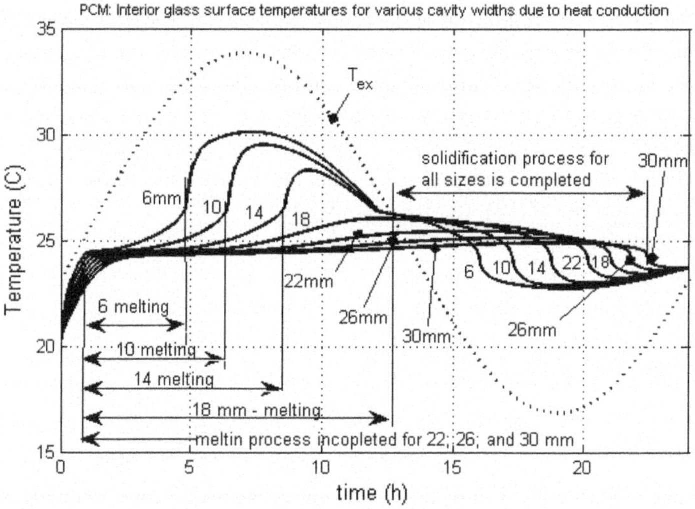

Fig. 5.16 Simulation results: temperature swing on the indoor glass surface for various cavity widths in 24-h period based on average $h_{ex} = 25$ W/m^{2}K (due to temperature differences only) (Original author's photo)

width regarding the temperature swing, average temperature, melting/solidification ratio and shown in Fig. 5.17.

A low *p*-value ($p < 0.05$) suggests that the slope is not zero, which in turn suggests that changes in the cavity width are associated with changes in the response variable. Also, there is a significant negative change in the average temperature ($with\ p = 0.01$)

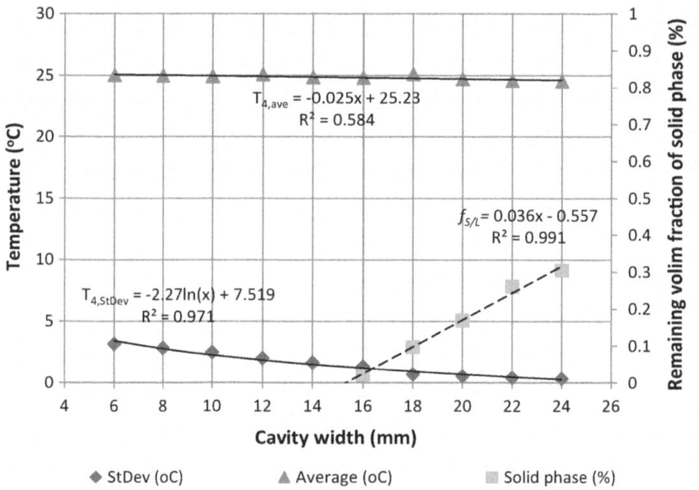

Fig. 5.17 Standardized temperature deviation as design criterion (Original author's photo)

and there is a significant reduction of the temperature variation ($p = 0.02$) with cavity with growth. It is obvious that the cavity widths above 16 mm keep some portion of the solid phase while the liquid phase completely solidifies for each cavity. Thus, as a recommended design criterion for the north-facing glazings and the condition with no sun radiation can be 16 mm cavity. The average temperature, temperature variation, and remaining solid phase for various cavity sizes have been described with a good fit by Eqs. (5.18), (5.19), and (5.20).

Average temperature

$$T_{4,\text{ave}} = -0.025x + 25.23 \tag{5.18}$$

Temperature variation

$$T_{4,\text{StDev}} = -2.27ln(x) + 7.519 \tag{5.19}$$

Remaining volume fraction of the solid phase after solidification process

$$f_{\text{S/L}} = 0.036x - 0.557 \tag{5.20}$$

where x is cavity width in mm.

Considering the other design criteria such as heat gain and the time that has little/no effect on the A/C load, similarly, the same conclusion can be drawn as it is depicted in Fig. 5.18.

Total unload time in a day increases with the increase of cavity width. The increase of the unload time can be well approximated by a logarithmic function with a pretty high coefficient of determination of $R^2 = 0.981$, which gives a very good fit for the

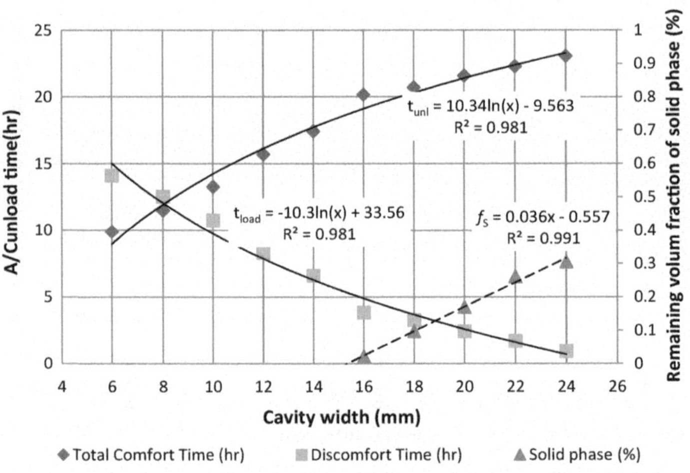

Fig. 5.18 Contributing versus noncontributing hours to A/C as design criterion (Original author's photo)

function. The unload time is growing fast with the increase of glazing gap up to 16 mm, and then the slope of the function is slightly reducing after 16 mm for each next cavity width. Conversely, the load time period decreases with the increase of cavity width per negative logarithmic function. The function has a very good fit as well with a high coefficient of determination of $R^2 = 0.981$. The decrease of the function is decreasing quickly for the cavity widths up to 16 mm, after which the slope of the function is slightly reducing for each next cavity width. Equations (5.21) and (5.22) are used to describe the relationship between cavity width and the hours reducing the stress on the A/C system.

Unload time

$$t_{\mathrm{unl}} = 10.34\ln(x) - 9.563 \tag{5.21}$$

Load time

$$t_{\mathrm{load}} = -10.3\ln(x) + 33.56 \tag{5.22}$$

The function slops suggest that a 16 mm cavity can be a threshold for the design of north-facing PCM glazings. Taking a closer look at what is going on with the PCM at 16 mm, it is noticeable that the remaining solid-phase ratio is increasing with the increase of the cavity width per a polynomial function with a high coefficient of determination ($R^2 = 0.99$). The relationship for the observed data is locally linear after 16 mm. Therefore, it is not reasonable to go with greater thicknesses because it will not be utilized all potentials the PCM in terms of accumulation of latent heat, i.e., melting of the complete amount of materials. Also, the increase of the thickness will linearly increase the material cost, weight and decrease the efficiency of glazing system design. As a recommendation for the glazings facing north per load/unload time criterion, again one can take a 16 mm gap as a recommended value to design a PCM-filled glazing (Fig. 5.19).

Latent heat (LH) charging time is increasing in with the cavity width per a polynomial until it reached a 16 mm gap. After this point, the component of the LH is rapidly decreasing with the cavity growth while the component of the sensible heat continues to increase. This phenomenon is caused by the increase of the solid-phase rate of the material between the glazings. Since the LH depends on the liquid-phase ratio, which is decreasing, the LH is decreasing as well per a linear function. On the other hand, SH charging time is linearly increased per linear function for all gap thicknesses. Again, this is caused by the increase in solid ratio, which stores sensible heat, in the PCM mixture between glazing panes. LH discharging time is increasing per linear function for all glazing gap widths. This phenomenon is caused by accumulated heat in the solid and liquid phases as sensible heat or latent heat. Since the temperature of the exterior environment is slowly decreasing below the liquidus temperature, the stored heat in the PCM as LH or SH is releasing to the environment at a constant rate for all thicknesses of the glazing gap. All these three components contribute to the energy saving in the buildings and therefore contribute to the occupant thermal comfort for certain periods of the day. A way of extending

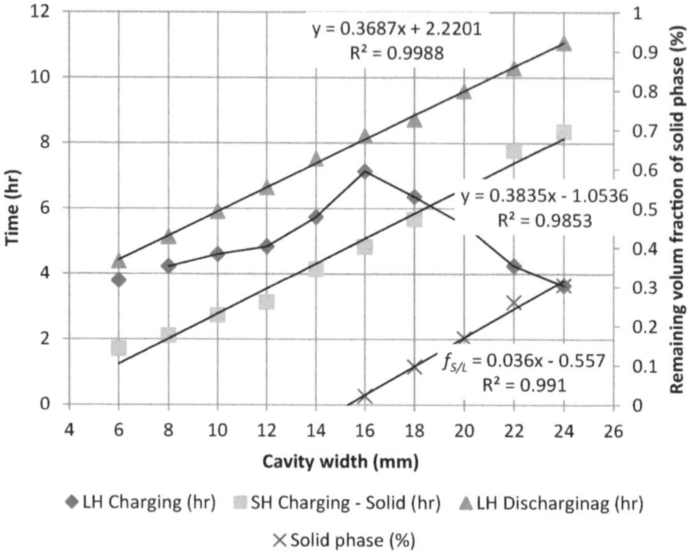

Fig. 5.19 Components of unload (noncontributing) hours (Original author's graph)

A/C unload time and contributing to energy-saving is the selection and use of the PCM with higher enthalpies and lower conductivities.

5.5.2.1 Impact of Convective Heat Transfer Coefficient to the Unload Time

Since the exterior glass surface is exposed to the different wind speed values, convective heat transfer coefficient on the surface is changing. Winds peed can vary from less than 0.2 m/s for calm weather, free convection conditions, to over 30 m/s for storm conditions. Accordingly, a nominal value of 29 W/m^{2}K (*corresponding to a* 6.7 m/s *wind*) is often used for glazing design [38]. The effect of convective heat transfer coefficient on the melting and solidification time is shown in Fig. 5.20. Three different convective coefficient values (15, 25, and 35 W/m^{2}K) were simulated for all cavity thicknesses with the aim of assessing the time needed for melting and solidification of the PCM as well as energy storage.

Variation of this parameter produces three nonlinear curves over cavity width represented by Eqs. (5.23), (5.24), and (5.25). Since the convective coefficient has a significant impact on the heat transfer to the exterior glazing, it is reflected on the melting of the PCM in the cavity as well. The remaining solid phase can be described by three linear functions corresponding to their coefficients, respectively. The more wind on the exterior surface the more heat transferred to the PCM and consequently

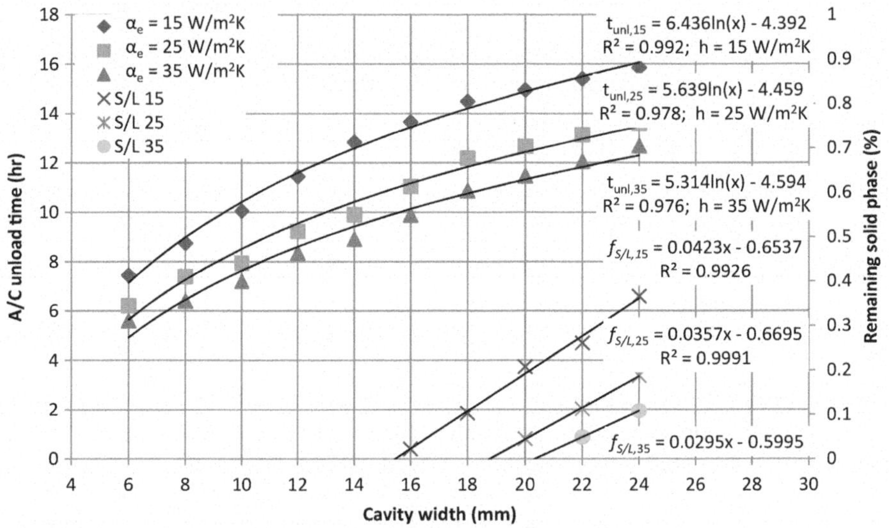

Fig. 5.20 Melting and solidification time with respect to the convective heat transfer coefficients (h_{ex} = 15, 25 and 35 W/m^{2}K) (Original author's photo)

the less solid phase per unit of time for the same cavity width and vice versa. The recommended cavity width is changeable and varies from 16 to 21 mm, which is described by Eqs. (5.26), (5.27), and (5.28). Finally, interdependences between the exterior convective coefficient and the remaining solid volume fraction were defined by exudation (5.29).

Unload time in case the convective heat transfer coefficient on the exterior side is 15 W/m^2

$$t_{\mathrm{unl},15} = 6.436\ln(x) - 4.392 \tag{5.23}$$

Unload time in case the convective heat transfer coefficient on the exterior side is 25 W/m^2

$$t_{\mathrm{unl},25} = 5.639\ln(x) - 4.459 \tag{5.24}$$

Unload time in case the convective heat transfer coefficient on the exterior side is 35 W/m^2

$$t_{\mathrm{unl},35} = 5.314\ln(x) - 4.594 \tag{5.25}$$

Remaining solid phase if h_{ex} = 15 W/m^2

$$f_{S/L,15} = 0.0423x - 0.6537 \tag{5.26}$$

Remaining solid phase if $h_{ex} = 25\ \mathrm{W/m^2}$

$$f_{S/L,25} = 0.0357x - 0.6695 \tag{5.27}$$

Remaining solid phase if $h_{ex} = 35\ \mathrm{W/m^2}$

$$f_{S/L,35} = 0.0295x - 0.5995 \tag{5.28}$$

where x is cavity width in mm.

Finally, the recommended cavity width as function of the convective heat transfer coefficient on the exterior glass surface can be described as follows:

$$x(h_{ex}) = 5.824 \ln(h_{ex}) - 0.238 \tag{5.29}$$

where h_{ex} is the convective heat transfer coefficient on the exterior glass surface, in $\mathrm{W/m^2K}$. Due to the radiation after the melting phase, an instantaneous jump occurs. The processes of solidification and melting are extending by the increase of cavity thickness. The authors do not recommend thickness beyond 24 mm of the cavity for the windows exposed to radiation. The increase of the cavity thickness is beneficial but limited by solid-/liquid-phase volume fraction. Thus, no more than 24 mm cavity is recommended for the PCM-filled window exposed to the radiation, meaning the south-oriented glazing. No more than 19 mm is recommended in glazing limited to the radiation, meaning the north-facing glazing [3].

5.6 Remarks

PCM-based glazing has the capacity to reduce interior temperature variance. The drawback is reduced transparency when the PCM is in solid-state, which qualifies the systems for building with reduced demand for light. Double-glazed system to be suitable for operating interchangeably in summer and winter modes, the PCM transition temperature has to be lower than the winter room temperature. Energy-efficient design of a PCM-based glazing system depends on used materials, solar radiation intensity, cavity size between glazing, wind speed on the exterior surface and building orientation.

Heat gain and temperature reduction to the interior can be achieved by a proper cavity thickness. Based on research results, it was observed that the increase of cavity size between glazing panes the radiative component of heat gain has a negative nonlinear relationship, while the temperature reduction on the interior surface has a negative linear relationship. This has a positive effect on A/C energy consumption reduction through its inactivity time increment as it was shown in Figs. 5.12 and 5.14. It was found that a 24-mm-cavity size is optimum for PCM glazings exposed to direct solar radiation, and it is limited by liquid volume fraction, which cannot solidify overnight. Increasing the cavity size over 24 mm will not contribute to the energy reduction since any amount beyond that portion will not get a chance to solidify overnight. For PCM glazings exposed to convection and diffuse radiation such as north-facing glazing, a maximum of 19 mm is recommended for spacing between glass panes [3].

References

1. Ismail KA, Salinas CT, Henriquez JR (2008) Comparison between PCM filled glass windows and absorbing gas filled windows. Energy Build 40(5):710–719
2. Silva T, Vicente R, Rodrigues F (2016) Literature review on the use of phase change materials in glazing and shading solutions. Renew Sustain Energy Rev 53(January):515–535
3. Duraković B, Torlak M (2017) Experimental and numerical study of a PCM window model as a thermal energy storage unit. Int J Law-Carbon Technol XII(3):272–280
4. Weinlaeder H, Koerner W, Heidenfelder M (2011) Monitoring results of an interior sun protection system with integrated latent heat storage. Energy Build 43(9):2468–2475
5. Mehling H (2005) Strategic project 'Innovative PCM-technology-results and future perspectives. In: 8th expert meeting and workshop. Kizkalesi, Turkey, 18–20 Apr 2005
6. Bayern ZAE (2004) Tätigkeitsbericht 2004, Annual Report 2004. ZAE Bayern
7. Dörken GmbH & Co. KG (2013) Phase change material—thermal storage mass for a comfortable interior climate, manufacturer information leaflet. [Online]. Available http://www.doerken.de/bvf-de/produkte/pcm/produkte/cool28.php
8. DELTA®-COOL 28 used in a glass facade application on a zero energy office building in Kempen, CH. Cosella-Dörken Products, Inc., [Online]. Available http://www.cosella-dorken.com/bvf-ca-en/projects/pcm/kempen.php. Accessed 19 Feb 2014
9. Alawadhi EM (2012) Using phase change materials in window shutter to reduce the solar heat gain. Energy Build 47:421–429
10. Soares N, Samagaio A, Vicente R, Costa J (2011) Numerical simulation of a PCM shutter for buildings space heating during the winter. In: Proceedings of WREC—world renewable energy congress. Linköping, Sweden
11. Silva T, Vicente R, Rodriguea F, Samagaio A, Cardoso C (2015) Development of a window shutter with phase change materials: full scale outdoor experimental approach. Energy Build 88:110–121
12. Buddhi D, Mishra H, Sharma A (2003) Thermal performance studies of a test cell having a PCM window in south direction. Annex 17, Indore, India
13. Ismail K, Henriquez JR (2001) Thermally effective windows with moving phase change material curtains. Appl Therm Eng 21(18):1909–1923
14. Merker O, Hepp F, Beck J, Fricke J (2002) A new solar shading system with phase change material (PCM). In: Proceedings of the world renewable energy congress VII. Cologne, Germany

15. GlassX (2013) GlassX homepage, Broschuere online. [Online]. Available http://glassx.ch/uploads/media/GLASSX_AG__products_e.pdf. Accessed 19 Feb 2014
16. Manz H, Egolf P, Suter P, Goetzberger A (1997) TIM-PCM, external wall system for solar space heating and daylight. Sol Energy 61(6):369–379
17. Bontemps A, Ahmad M, Johannès K, Sallée H (2011) Experimental and modelling study of twin cells with latent heat storage walls. Energy Build 43(9):2456–2461
18. Cosella-Dörken Products Inc. (2012) DELTA®-COOL 28 used in a glass facade application on a zero energy office building in Kempen, CH. [Online]. Available http://www.cosella-dorken.com/bvf-ca-en/projects/pcm/kempen.php. Accessed 19 Feb 2014
19. Goia F, Perino M, Serra V, Zanghirella F (2010) Experimental assessment of the thermal behaviour of a PCM grazing. In: IAQVEC 2010—the 7th international conference on indoor air quality, ventilation and energy conservation in buildings Syracuse. Syracuse, New York, USA
20. Goia F, Perino M, Haase M (2012) A numerical model to evaluate the thermal behaviour of PCM glazing systems. In: IBPC 2012—the 5th international building physics conference. Kyoto, Japan
21. Goia F, Perino M, Serra V (2013) Improving thermal comfort conditions by means of PCM glazing systems. Energy Build 60(May):442–452
22. Weinläder H, Beck A, Fricke J (2005) PCM-facade-panel for daylighting and room heating. Sol Energy 78(2):177–186
23. Gowreesunker BL, Stankovic SB, Tassou SA, Kyriacou PA (2013) Experimental and numerical investigations of the optical and thermal aspects of a PCM-glazed unit. Energy Build 61:239–249
24. Li D, Ma T, Liu C, Zheng Y, Wang Z, Liu X (2016) Thermal performance of a PCM-filled double glazing unit with different optical properties of phase change material. Energy Build 119:143–152
25. Li D, Li Z, Zheng Y, Liu C, Hussein AK, Liu X (2016) Thermal performance of a PCM-filled double-glazing unit with different thermophysical parameters of PCM. Sol Energy 133:207–220
26. Goia F, Zinzi M, Carnielo E, Serrad V (2015) Spectral and angular solar properties of a PCM-filled double glazing unit. Energy Build 87(January):302–312
27. Li S, Zhou Y, Zhong K, Zhang X, Jin X (2013) Thermal analysis of PCM-filled glass windows in hot summer and cold winter area. Int J Low Carbon Technol (2013)
28. Zhong K, Li S, Sun G, Zheng J, Zhang X (2015) Simulation study on effects of phase change material thermal parameters on dynamic heat transfer performance of PCM-filled glass window. In: 6th international building physics conference, IBPC 2015
29. Li S, Zhong K, Zhou Y, Zhang X (2014) Comparative study on the dynamic heat transfer characteristics of PCM-filled glass window and hollow glass window. Energy Build 85(December):483–492
30. Grynning S, Goia F, Time B (2015) Dynamic thermal performance of a PCM window system: characterization using large scale measurements. Energy Procedia 78:85–90
31. Durakovic B, Yildiz G, Yahia ME (2020) Comparative performance evaluation of conventional and renewable thermal insulation materials used in building envelops. Tehnicki vjesnik—Technical Gazette 27(1) (in press)
32. Zhong K, Li S, Sun G, Li S, Zhang X (2015) Simulation study on dynamic heat transfer performance of PCM-filled glass window with different thermophysical parameters of phase change material. Energy Build 106(November):87–95
33. Durakovic B (2017) Design of experiments application, concepts, examples: state of the art. Periodicals Eng Nat Sci 5(3):421–439

34. Duraković B, Torlak M (2017) Simulation and experimental validation of phase change material and water used as heat storage medium in window applications. J Mat Environ Sci VIII(5):1837–1846. ISSN: 2028-2508. Copyright © 2017
35. Durakovic B, Torlak M (2017) Experimental and numerical study of a PCM window model as a thermal energy storage unit. Int J Low-Carbon Technol 12(3):272–280
36. Ismail KA, Salinas CT, Henriquez JR (2009) A comparative study of naturally ventilated and gas filled windows for hot climates. Energy Convers Manag 50(7):1691–1703
37. Duraković B, Mešetović S (2019) Thermal performances of glazed energy storage systems with various storage materials: an experimental study. Sustain Cities Soc 45(February):422–430
38. ASHRAE Handbook of Fundamentals USA (2009) American Society of Heating, Refrigerating and Air-Conditioning Engineers, Inc.

Chapter 6
PCMs in Separate Heat Storage Modules

Heat storage in separate storage modules requires active components such as control system, fans, and pumps to move the air and heat transfer fluid (HTF). The main advantage of this system is the accessibility to the stored heat when it is required. Various temperatures can be used depending on the area of application. Since the occupant comfort temperature requires a range between 20 and 27 °C depending on cooling/heating season, the storage temperatures of the PCM are favored in the range of 0 and 40 °C (for cooling/heating) [1]. The exception of that is hot water and heating water provision where the temperature range is in-between 50 and 60 °C. Therefore, PCM in separate heat storage modules is a concept combined with building services technology. For example, the heat from the sun is absorbed by solar collector and transferred by HTF to the storage tank filled with the PCM. Another HTF circulating loop transfers the heat from the tank to the heated room on the demand (see Fig. 6.1). Also, Fig. 6.1 at the same time provides a rough classification of research published on PCMs in separate heat storage modules for building applications. It can be classified into three classes: PCM-based heat collectors, PCM-based heat bank, and PCM-based heaters.

The heat transfer fluid (HTF) enters the heat collector at the bottom and absorbs incoming solar radiation over the collector by increasing its temperature. Increased temperature fluid enters the heat bank with the PCM, and the heat from the HTF is stored in the bank by melting the PCM. Another water loop circulates between the room and the storage tank by taking the heat from the tank and heating the interior. Storage modules for building application are based mainly on three different concepts (see Fig. 6.2) [2].

- The PCM is placed in a storage tank, and the HTF flows through *channels* into a heat exchanger.
- The PCM is *macroencapsulated* in PCM modules that are located in the storage container—the HTF flows around the capsules.
- The PCM is a component of the HTF and increases its capacity to store the heat—called "*PCM slurry.*" Thus, it can be pumped to any given location in the system in order to release or absorb heat directly.

© Springer Nature Switzerland AG 2020

B. Duraković, *PCM-Based Building Envelope Systems*, Green Energy and Technology, https://doi.org/10.1007/978-3-030-38335-0_6

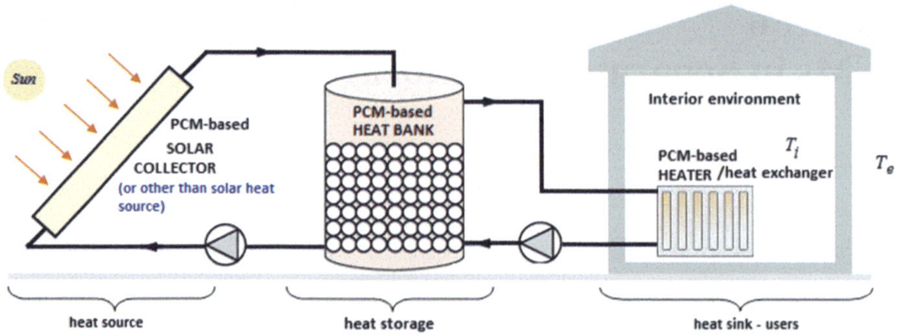

Fig. 6.1 Classification of PCMs in separate heat storage modules in a typical heating/cooling system

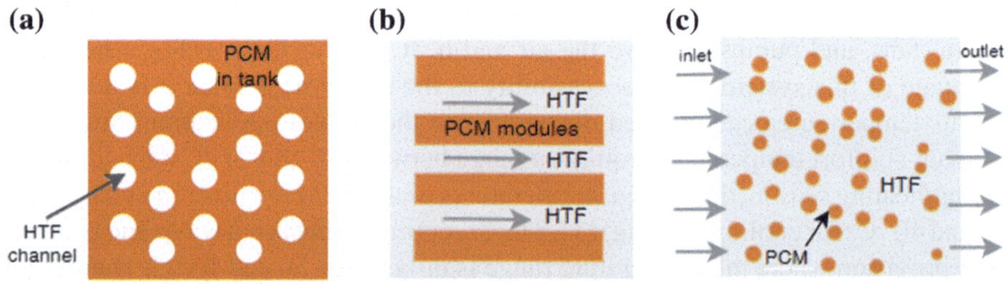

Fig. 6.2 Storage concepts for PCM integration in heat modules: **a** PCM in storage tank with channels; **b** PCM in modules; **c** PCM slurry

The HTFs in the first two concepts can be used as air, water, or other fluids, whereas the third concept is only suitable for liquids [3].

Basically, a number of studies have been done on the integration of PCMs in solar collectors, heat banks, as well as indoor heat exchanger. Heat modulation based on PCMs has been tested and installed in different parts of the building as well as outside of building over decades. Wide usage of PCMs has been decelerated due to high initial cost, corrosiveness that is present in some inorganic PCMs, and loss of phase change capability. Paraffin waxes are low-priced but have low thermal conductivity, and their application demand bigger surface area [4]. Furthermore, paraffinic hydrocarbons have high flammability factors and put the building on danger [5]. As a result, research is now focused on fatty acids or inorganic salt hydrates PCMs that have larger energy storage. The integration of hydrated salts demands the usage of thickening agents and nucleating, due to supercooling and phase segregation [4]. PCM in separate heat storage modules can be divided in three major groups (Fig. 6.3).

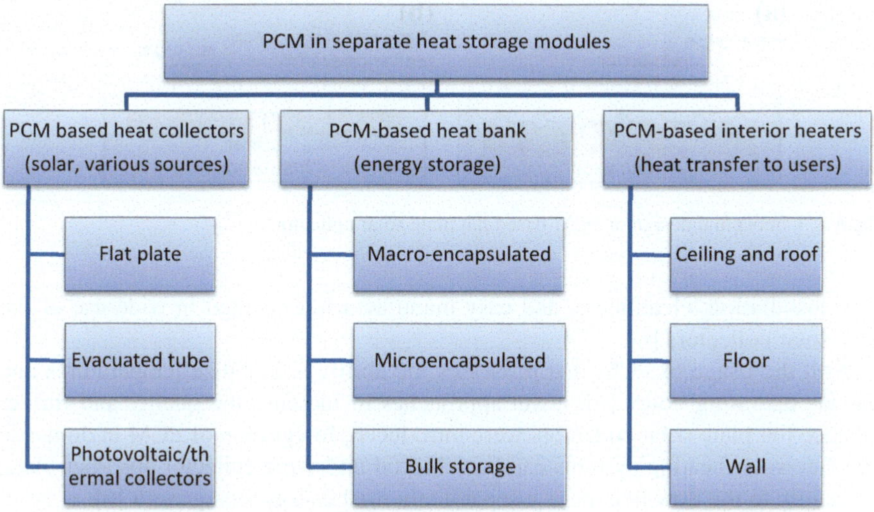

Fig. 6.3 Classification of PCMs in separate heat storage modules

6.1 PCM-Based Solar Collectors

Solar collector, as an integral part of the solar thermal system, absorbs the solar radiation and further via heat transfer fluid (HTF) transport to the heat storage bank or users. Various types of solar collectors are used such as flat plate, evacuated tube, and photovoltaic/thermal collectors.

In order to improve the thermal performances of solar collectors, various enhancements have been made. In recent times, new technological advancements allowed integration of the phase change materials into solar collectors using different methods. Many numerical and experimental studies have been conducted in order to explore the integration of PCMs in solar collectors.

PCM can be considered as a separate entity for latent heat energy storage and can be placed between the source of energy and user. However, PCM can also be integrated into a solar flat plate, evacuated tube, and photovoltaic/thermal collectors.

6.2 Flat Plate Solar Collector

The most common technology for converting solar energy into heat in building technology is flat plate solar collector. The flat plate solar collector by a greenhouse effect collects solar energy and delivers necessary hot water into buildings, at low cost and relatively easy installation. Due to the possibilities to operate in low-temperature ranges, flat plate collectors are commonly applied in industrial and domestic spaces.

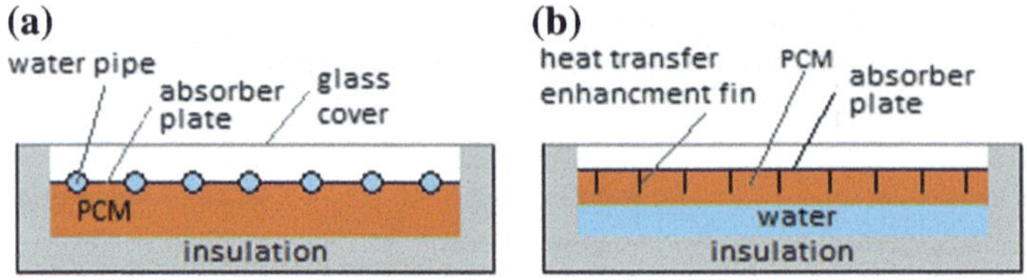

Fig. 6.4 Conceptual design of PCM-based flat plate solar collector

Easy installation, cleanliness, and easy manufacturing resulted in wide use of flat plate solar collectors [6].

With the usage of PCM that possesses the ability to stabilize temperatures and extends operating hours [6], novel approaches to increase the quality and further enhance flat plate solar collectors were introduced. Integration of PCM in domestic flat plate solar heating systems can be achieved under the collector absorber plate, concentric to the flow line, or as a separate thermal energy storage unit [6].

Due to low cost, easy manufacturing and satisfactory improvements of thermal conditions, flat plate collectors are mostly investigated solar heating collectors. The integration of PCM into this type of solar collectors is represented in Fig. 6.4.

PCMs improved thermal stability and operation hours [7] of solar collectors. Insulation material [8] is used to reduce energy dissipation to environment from the collector. The results of various studies of the integration of PCM into flat plate collectors showed that backup time is mostly proportional to the thermal efficiency [9, 10]. It is also reported that discharging takes place faster than charging. Mostly utilized PCM was paraffin that increased thermal stability and heat transfer. The key results of different integration of PCMs into solar collectors are shown in Table 6.1.

6.3 Evacuated Tube Solar Collectors

Evacuated tube collectors are converting sun energy into heat in a solar water heating system. They are constructed of one or several rows of tubes connected to the manifold. Each tube has an outer thick glass tube with a thinner inside glass tube coated with solar energy absorbent material. Due to the effect of vacuum inside the tubes, evacuated tube collectors more efficiently obstruct heat loss compared to flat plate collectors. The single-walled glass and the Dewar tube are two major groups into which evacuated solar tubes are divided, while there are many more categorizations inside these two [14]. Energy converted by evacuated tube collectors can be used for both domestic and commercial heating and air-conditioning.

In the inner part of each glass tube, there is an aluminum or copper fin absorber fastened to an inner tube copper heat pipe. The absorber is coated with the layer that absorbs and transmits the heat to the fluid inside the pipe. From the internal heat

Table 6.1 Experimental and numerical results for flat plate efficiencies

Type of solar collector	PCM	Type of study	Thermal efficiency improvements	References
Flat plate/alcohol	Tricosane (paraffin wax 116), water, and sodium acetate	Experimental	Charge efficiency of 73%, discharge efficiency of 81%	[11]
Experimental	Paraffin wax	Experimental	Ranges around from 45 to 54% for a clear day in January and February and partially cloudy day in March	[12]
Flat plate/water	Paraffin wax	Experimental	PCM provides thermal efficiency of around 52% and 38 °C water temperature	[10]
Flat plate/water	Microencapsulated PCM (n-eicosane)	Numerical	Instantaneous efficiency in the range 5–10%; converted heat by the slurry PCM-based system in wintertime is 20–40% higher than of a conventional water-based solar collector	[13]
Flat plate/water	Paraffin/water	Experimental and numerical	Energy efficiency between 25–35%	[9]

transfer fluid, the copper pipe transfers heat via convection. In later process, the heat is transferred to a copper manifold and the header tank.

- **Evacuated tube**—absorbs solar energy that is later converted to heat. An insulation from heat loss is created by the vacuum between the two layers of glass.
- **Heat transfer fin**—improves the heat transfer to heat pipe.
- **Heat pipe**—heat transfer from evacuated tube to the manifold.
- **Manifold**—insulated frame containing the header pipe. The header consists of a pair of copper pipes with sockets for connecting heat pipes.
- **Mounting frame**—a frame for installing evacuated tube collector.

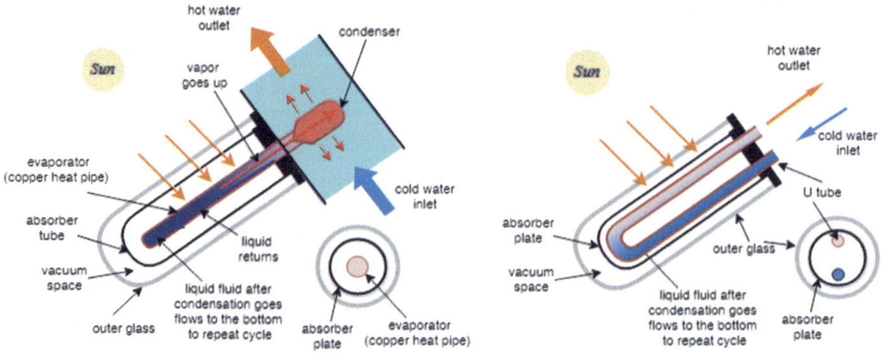

Fig. 6.5 Conceptual design of heat piper. **a** evaporating tube; **b** U-tube

Due to the vacuum insulation and surface coating, evacuated tube solar collectors (ETSCs) had a lower heat loss compared to a flat plate collector, while the outlet temperature of fluid is higher [6].

The first example of integration of PCM into evacuated tube collector was published in 2006 by Riffat et al. [15]. This experimental study reported integration of energy storage materials, paraffin wax and water, into cylindrical evacuated solar collector, and investigated the possibilities of providing continuous hot water. Subsequently, the integration of PCMs into evacuated tube collector was more examined. M. M. A. Khan et al. divided evacuated tube collectors with PCMs into two categories [6]. PCM integrated into manifold in direct contact with part of the heat pipe is the first group.

This PCM application is possible in heat pipe evacuated tube collector type. The theoretical study by Naghavi et al. from 2015, for charging and discharging, used water as the collector fluid [16]. The process of transferring thermal energy from the condenser section of heat pipe to PCM and fluid that flows in the transverse direction with the heat pipe is represented in Fig. 6.5.

In an experimental investigation study by Mehla and Yadav from 2015 [17], the latent thermal energy storage was integrated into solar collector manifold while the working fluid that was transferring solar gain to heat was used water.

Integration of PCM into ETSC can be achieved in the tube structure of the collector, while the application can have many arrangements. Some of the numerical and experimental studies of integration of PCM into ETSL are summarized in Table 6.2.

Contrary to flat plate collectors, there are no significant breakthrough studies of the integration of PCM into ETSC collectors in the literature [6]. This can be due to the complexity of ETSC collectors. Different methods of integration of PCM into ETSC are represented in Fig. 6.6.

Table 6.2 Research results based on numerical and experimental studies for evacuated tube

PCM	Type of study	Result	References
Paraffin wax	Numerical	Improved performance compared to the conventional system (water flow rate higher than 55 lph); The proposed system sensitivity of the efficiency to the draw-off water flow rate less than the baseline system	[16]
Acetamide	Experimental	With different configuration PCM efficiency ranges from 14.43 to 17.65%	[17]
Tritriacontane, Erythritol	Experimental	When compared to standard solar water heaters, efficiency improved by 26% for the normal operation and 66% for the stagnation mode	[18]
Paraffin wax	Experimental	Useful heat gain from the paraffin-integrated ETC/S system increased by 45–79%, in accordance with the heating medium mass flow rate during the discharge cycle	[19]

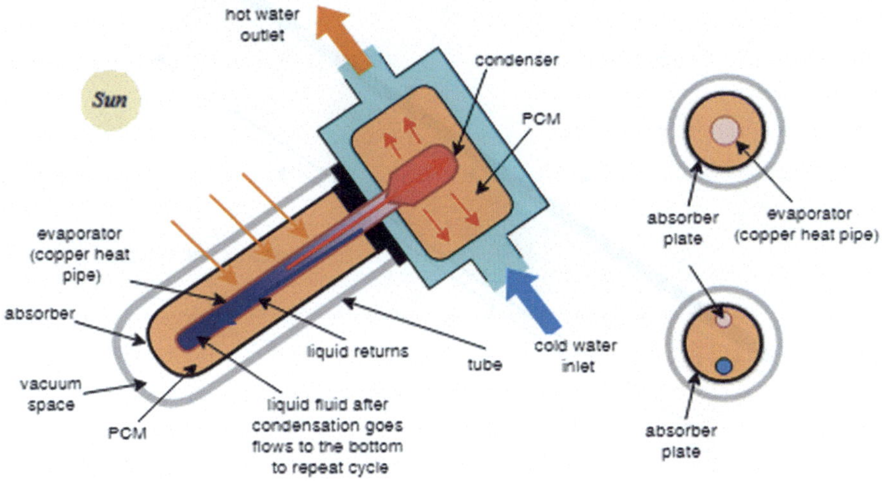

Fig. 6.6 Conceptual design of PCM-based heat piper

6.4 Photovoltaic/Thermal Collectors

The solar spectrum consists of about 50% infrared light, 40% visible light, and 10% ultraviolet light. In the photovoltaic module (PV), the ultraviolet and visible solar energy components are converted to electricity, while infrared components are transformed into heat. This process increases the cell temperature and reduces electrical efficiency [20]. Extraction of accumulated excessive heat in PV modules up to 80 °C [21] was done applying a photovoltaic/thermal (PV/T) concept. This concept utilizes the usage of the thermal collector with PV panel as it is shown in Fig. 6.7.

The integration of PCM into a PV system in buildings was firstly examined in 2003 by Huang et al. [22] that used PCM to moderate temperature rise by experiments and numerical simulations. The results and key data for some of the studies are represented in Table 6.3.

Active heat removal technique within PV performance combined with organic PCM and some of the inorganic PCMs such as fatty acids gave the best results compared to many other methods, such as natural convective ways [6]. Different concepts of integrating PCMs in flat plate solar collectors and combined with PV cells are shown in Figs. 6.4 and 6.7.

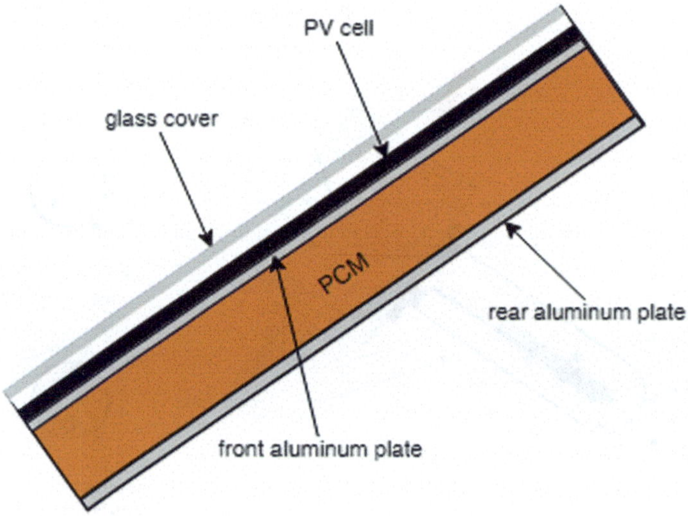

Fig. 6.7 PCM-based plate collector used with for PV cooling

Table 6.3 Experimental and numerical results for photovoltaic/thermal collectors

PCM	Type of study	Result	References
Paraffin wax	Numerical	9% increase in PV performance and an average water temperature rise of 20 °C	[23]
Paraffin wax	Numerical	10.7% increase in overall efficiency when compared to "no phase change material" mode	[24]
Pure salt hydrate ($CaCl_2{\cdot}6H_2O$) and eutectic mixture of fatty acids (capric–palmitic acid)	Experimental and numerical	Lower PV temperatures and higher power are achieved with $CaCl_2{\cdot}6H_2O$; effective only in warm and stable climates	[25]

6.5 Heat Bank

The development of renewable energy sources provided numerous studies on the integration of solar energy into existing energy systems. To effectively utilize solar energy and solve the problems of the intermittent nature of solar energy, storage systems must be developed. In the development of more effective storage systems, PCMs have shown promising results.

The thermal energy can be accumulated into heat banks, devices that can be sensible or latent heat storage or combination of both. The stored materials increase in temperature while the energy is stored in sensible heat storage, whereas in the case of latent heat storage systems the energy becomes useful when the substance changes its phases.

The latent heat thermal storages consist of three main components:

- Phase change material
- Container for the PCM encapsulation
- Heat exchange surface that transfers the heat from heat source to PCM and to the heat sink [26].

The incorporation of PCM into storage tanks was examined by analyzing the influence of PCMs integrated solar storage tanks on the domestic hot water system performance [27]. In a typical solar heating system, the heat bank can be a tank completely filled with PCM that contains channel for HTF (or bulk storage) [28] or has integrated PCM capsules or modules inside the tank for heat storage [29, 30] as it is shown in Fig. 6.8.

The advantages of these storage systems are the possibility of storing a high amount of excess solar energy and releasing it for long period water heating and reducing heat loss from the system. Disadvantages of PCM solar tanks within are

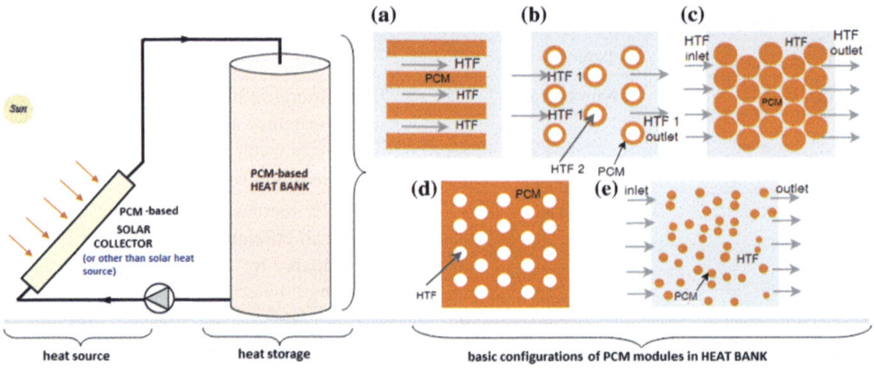

Fig. 6.8 Configuration of PCM modules in heat bank: **a** flat plate; **b** shell and tube with cross-flow; **c** sphere packed bed; **d** bulk storage in tank with tubes; **e** slurry

the high fluctuation of the temperature of hot water and high cost and thermal stratification [31]. Referring to Fig. 6.8, the way of encapsulation of PCMs in the heat bank can be categorized in three major groups: macroencapsulation, bulk storage, and microencapsulation.

6.5.1 Macroencapsulation

Macroencapsulation represents the most common type for the containment of PCM. One of the advantages of this system is the possible applicability to liquid and air as heat transfer fluids. Macrocapsules can have different shapes, from rectangular to sphere. Some of the advantages of microencapsulation are increasing the rate of heat transfer and providing a self-supporting structure for PCM [26].

Common PCM macroencapsulated container geometry

The most important factor that has a direct influence on the heat transfer characteristics in the PCM and ultimately affects the melting time is PCM heat bank. This is especially important for macroencapsulated modules for modular storage concepts. The following factors to be considered are:

- PCM container geometry
- The container thermal and geometric parameters required for a given amount of PCM.

Most published papers dealing with latent heat thermal storage reveals that PCM containers are typically designed as long thin heat pipes [32], cylindrical containers [33, 34], shell-and-tube model, or rectangular containers [30, 35, 36]. Figure 6.9 shows the most commonly used container geometries.

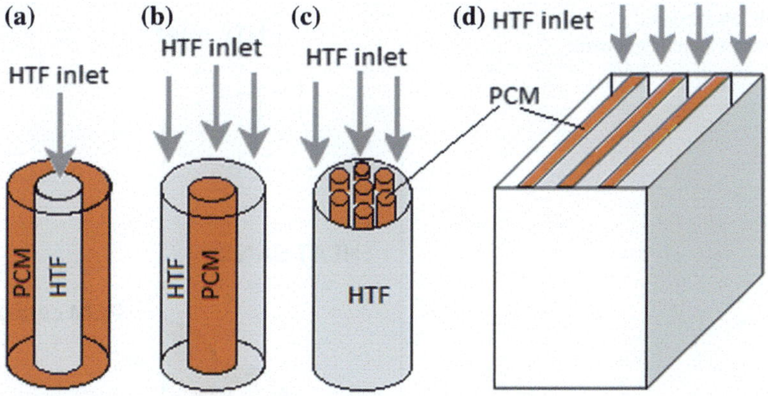

Fig. 6.9 PCM containers: **a** shell and tube—pipe module; **b** and **c** shell and tube—cylinder module; **d** flat plate module

The literature survey revealed that the shell-and-tube system is the most analyzed system. The fact is that pipes are most in use in engineering and that heat loss from shell-and-tube system is minimal.

PCM sphere packed bed heat bank

PCM beds are designed in accordance with operating temperature limits, PCM melting temperature, thermal load, the configuration of the bed and PCM latent heat. PCM beds can consist of microencapsulated PCM, together with the container and heat transfer fluid that transfers heat through the bed voids [26]. As the hot heat transfer fluid circulates during the charging mode, the capsulated PCM absorbs latent heat and melts. When cool heat transfer fluid circulates inside the tanks during the discharging mode, it freezes the PCM. In order to have efficient heat transfer between the fluid and packed bed, size, shape, material elements, and the geometry of the container must be carefully determined. Heat transfer is also affected by fluid properties, thermal state of the bed, inlet temperature of the fluid, and convective coefficient of heat transfer. An example of an LHTS unit with PCM capsules is shown in Fig. 6.10.

Hydrated salts as PCMs have phase change temperature that is lower and are applied into buildings and as part of solar water heating, while metals and molten salts have high phase change temperature and are mostly integrated into solar power generation and as an industrial heat waste recovery [37]. Molten salts as inorganic PCMs can also be incorporated into TES bank; however, due to low thermal conductivity and leakage, molten salts need additional modifications prior to use. Wu et al. [38] examined thermal properties of storage with molten-salt packed bed and concluded that effective heat release efficiency can be improved by the increase of the phase change temperature, reduction of entrance velocity of molten salt, and reduction of PCM capsules diameter.

PCMs as the energy storage materials possess high heat storage capacity and absorb or release energy at a constant temperature during phase change. This is

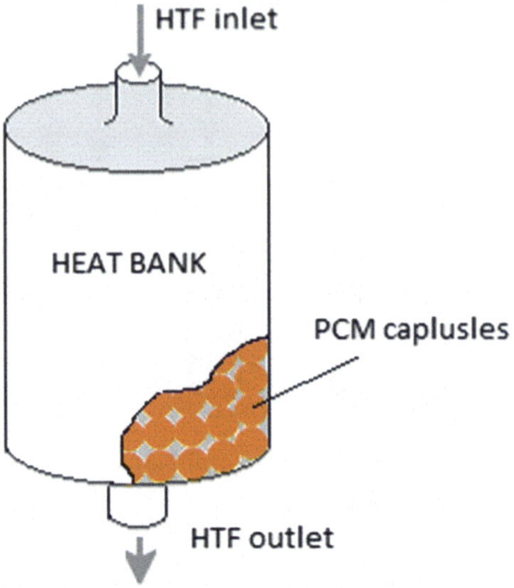

Fig. 6.10 Packed bed macroencapsulated heat bank

why they are considered the best solution in storing energy for heating and cooling systems. In accordance with the physical and chemical characteristics of PCM, the design of the storage unit is decided. If the energy density of PCM is higher, then the smaller storage volume is necessary [26]. The use of PCM as material for heat storage is decided on their thermal, chemical, and economic properties.

6.5.2 Bulk Storage

Due to similar design to existing tanks for energy storage, bulk storage is usually referred to as the "tank heat exchanger" for PCMs [39] (see Fig. 6.8d). It requires a more extensive heat transfer due to the fact that PCMs have a high heat storage density and lower PCM thermal conductivity in comparison to micro- and macroencapsulation [26]. However, with the integration of fins, direct contact with the heat exchanger surface, and improving PCM by the addition of particles, PCM can have higher conductivity properties [40].

6.5.3 Microencapsulation

Microencapsulation of a phase change material represents a method in which PCM can change phase without affecting the environment where used. In this case, a large number of PCM particles are contained in a matrix that needs to have high thermal conductivity. The heat transfer can be decreased due to the system that forces conduction instead of convection. Microencapsulation is rarely used due to increased costs.

6.6 Thermal Conductivity Enhancement Techniques

Although phase change materials have high heat storage capacity and during phase change maintain a constant temperature, they also have a low thermal conductivity that causes low heat transfer and low heat storage and release rate. However, these disadvantages can be improved by various enhancement techniques, such as the introduction of thermally conductive metallic and carbon-based nanoparticles, encapsulation of PCM, and expanded graphite [41]. The classification is shown graphically in Fig. 6.11.

Carbon-based and metallic nanoparticles have high thermal conductivity and are usually integrated into PCMs to improve thermal conductivity. However, carbon-based nanoparticles such as carbon fiber, carbon nanotubes, and graphene nanoparticles have better stability and dispersion in PCM than metal-based nanoparticles. Due to its high thermal conductivity and stable thermal and physical characteristics, metallic foam is also used as a means to enhance PCMs.

Thermal conductivity improvement can be achieved by the integration of expanded graphite, as it has high thermal conductivity, low density, and excellent chemical and physical properties. Oxidation of graphite in the presence of nitric and

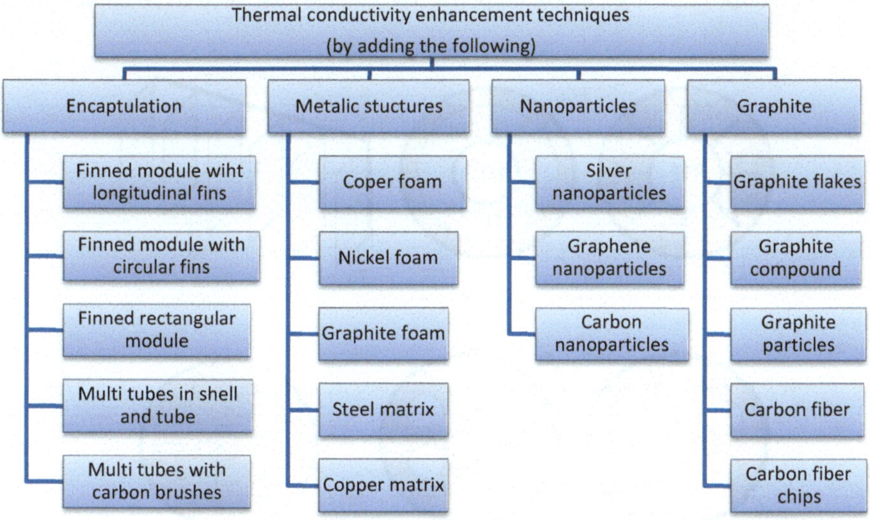

Fig. 6.11 PCM thermal conductivity enhancement techniques

sulfuric acid is the process of preparing expanded graphite [42]. After drying in vacuum oven and heating in a furnace, it gets expanded. Molten PCMs are later mixed with the expanded graphite, and composite PCMs are created.

Another method for improving PCM thermal conductivity is by encapsulation in the shell of organic material with high thermal conductivity. Encapsulation is mostly achieved using polyurea, urea-formaldehyde, and melamine-formaldehyde resin as shells [41].

6.6.1 Finned Tubes Enhancement of PCM

Low thermal conductivity leads to slow down charging and discharging rates of LHTES units. Since most PCMs have low thermal conductivity, heat transfer enhancement techniques are required. Studies on heat transfer enhancement techniques in PCMs include finned tubes of different configurations (circular and longitudinal) embedded in the phase change material to extend the heat transfer surface. It is a simple solution, easy to fabricate, and less expensive than improving the thermal conductivity of the PCM by high conductive materials or using intermediate heat transfer medium and employing multiple PCMs. Some of the heat transfer enhancement techniques for PCMs are shown in Fig. 6.12.

Due to high storage capacity and maintaining a constant temperature during a phase change, PCMs are widely used in heat and cooling systems. However, due to low thermal conductivity, the practical application represents a challenging process. The addition of high thermal conductivity materials, such as metallic foams

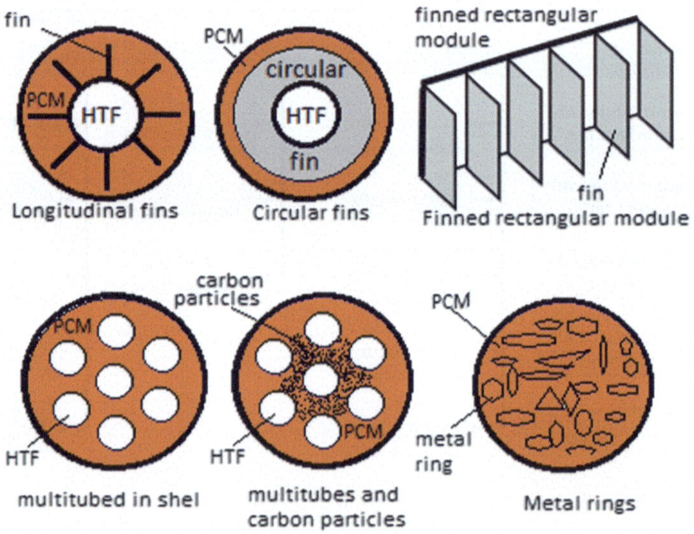

Fig. 6.12 Heat transfer enhancement techniques for macroencapsulated PCMs

and expanded graphite, results in improved thermal conductivity of PCMs. On the other side, the integration of different additives can have some negative impacts as well. Metal-based materials can lead to poor stability, while nanoparticles have non-uniform dispersion. Metallic foams have low density and good stability, and their performance is better than nanoparticles. Encapsulation of PCMs results with high thermal conductivity, and due to the obstruction of direct contact between PCM and surroundings, it prolongs usage and solves leakage.

Different methods of enhancing PCMs can also result in disadvantages that leave the space for future exploration and further enhancements of PCMs.

6.6.2 Nano-composite Enhanced PCM-Based Solar Collectors

In order to improve the low thermal conductivity of PCM nanotechnology is applied. The first combination of PCM with nanoparticles to improve the thermal storage performance was examined in 2005 by Ceng et al. [43]. The further step was examining the possibilities of integrating nanoparticles into solar thermal applications in 2009 by Shin and Banerjee [44]. After this, various studies to test further enhancements of nanocomposite PCM-based solar collectors were conducted.

Figure 6.13 shows the scheme of a flat plate collector with PCM elaborated in the experimental investigation by Saw et al. in 2013 [45]. The experiment was obtained using a 1.44-m^{2}-flat plate collector with nanocomposite PCM for water heating. PCM was paraffin wax with 20–nm-copper nanoparticles. The result indicated that the flat plate efficiency was improved by 6.9 and 8.4% with PCM and nanocomposite PCM.

The study conducted by Ma et al. [46] showed the difference in using PCM and nano-enhanced PCM integrated into the ceiling ventilation system. Within the air heating, PVT and ceiling were considered PCM layers with air gap between them as shown in Fig. 6.7. PCM RT24 was used as the base fluid while the PCM was mixed with 10% copper nanoparticles. The effect of nanoparticles was investigated analytically. It was concluded that the 8.3% more heat charged and 25.1% more heat discharged with PCM with nanoparticles compared to PCM without the addition. Above mentioned and many other studies showed that adding nanoparticles into

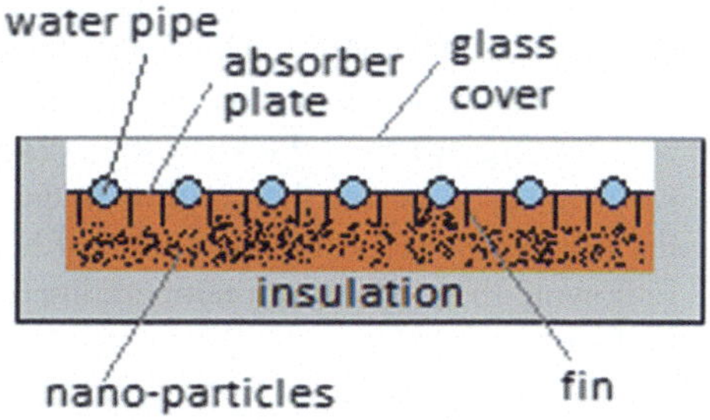

Fig. 6.13 PCM-based plate collector enhanced with nanoparticles and fins

PCM increases the thermal conductivity by directly enhancing the efficiency of the solar collector [6].

6.7 Interior PCM-Based Heat Exchangers

PCM-based heat exchangers within interior environment can be combined with a passive or active system. Depending on the system, it can be integrated into ceilings/roofs, floors, walls, or ventilation channels. As discussed earlier, passive systems do not use additional energy or mechanical devices to run PCM-based heating or cooling system. The system is regulated by temperature and air density fluctuations. Active systems need mechanical equipment to help achieve charging and discharging of the PCM thermal energy storage module. In active systems, PCM can be integrated into different parts of building units, such as storages, HVAC systems; or in case of solar cooling, it can be used as a heat/cold storage tank.

6.7.1 PCM Packed Bed in Ceiling Plenum and Walls

The free cooling system is dependent on the accumulated ambient cold that is later used during daytime. There are three systems for storing thermal energy for free cooling: sensible heat storage of heat accumulated by changing the internal energy, latent heat storage achieved by changing phases of the stored thermal energy, or the combination of these two principles [47, 48]. In regards to high-energy storage density and isothermal storage process, latent heat thermal energy storage (LHTES) is preferred.

One of the characteristics of free cooling is that mechanical equipment, such as fans, is used in order to charge or discharge the heat from the storage unit [49]. It is based on the principle of storing cold in the cases when the ambient temperature is lower than the indoor temperature. Once stored, it can be released in the room on demand, using electric fans [50].

Integration of PCM in free cooling has two operation modes shown in Fig. 6.14:

- Charging process—solidification of PCM: When the ambient temperature is lower than the interior temperature, the exterior temperature flows due to fans and absorbs heat from PCM. This starts the solidification process, until near thermal equilibrium between interior and exterior temperatures is achieved.
- Discharging process—PCM melting: This process is occurring when the interior temperature is above the comfort range. In this case, hot air is flowing next to the storage unit causing absorption of heat by the solid PCM that starts melting. Following this, the interior temperature becomes lower, and cooler air is delivered in the interior [51].

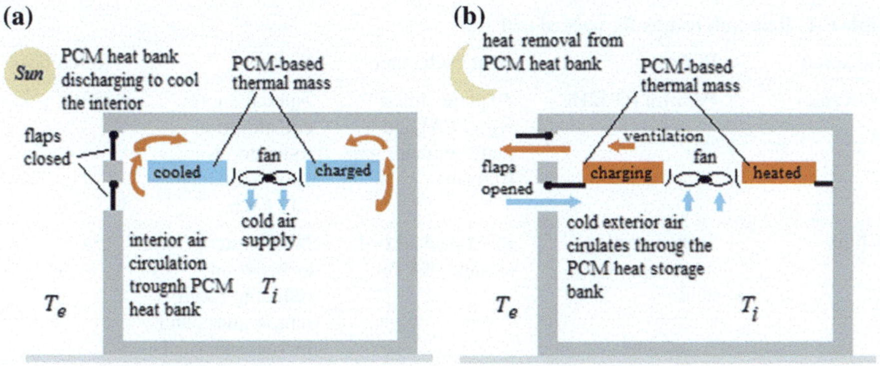

Fig. 6.14 PCM packed bed in ceiling plenum for free cooling: **a** Discharging to cool interior during the day—interior excess heat storage within thermal mass; **b** charging during the night—heat removal from thermal mass by ventilation to the exterior

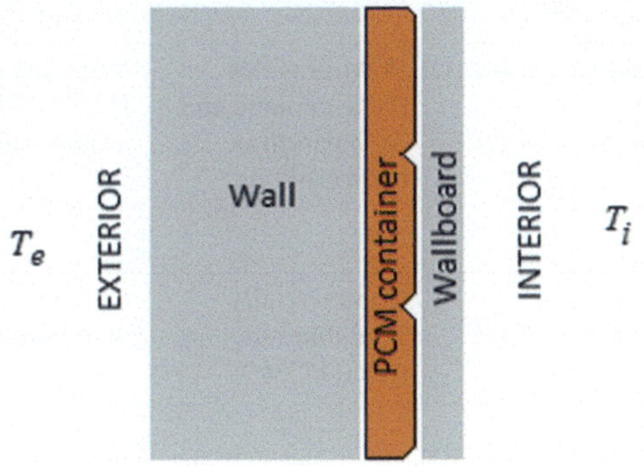

Fig. 6.15 PCM container in building wall

Melting and solidification of PCM are affected by parameters such as airflow rate, interior and exterior temperature, PCM melting temperature, thermo-physical, and encapsulation properties of PCM. An example of integration of a PCM module within building wall is shown in Fig. 6.15.

The PCM is encapsulated in thin polymer bags configured in sheets laminated with aluminum foils. The containers are placed between the wall and wallboard on the interior side of the wall [52].

Free cooling was examined in many studies, and some of the keynotes and results are presented in Table 6.4.

6.7.2 PCM Packed Bed in Floors and Ventilation Channel

In the last decades, the use of solar energy was widely examined; however, it was not until the last few years that the solar cooling systems based on adsorption and

Table 6.4 Research results for packed bed

Location	PCM	Heat exchanger	Results	References
Slovenia	Paraffin RT22HC	30 plates filled with PCM and plates with air gap (0.8 cm)	Annual energy consumption reduction for approximately 142 kWh	[53]
China	2000 capsules containing fatty acid	PCM packed bed storage (NVP)	NVP system effective in reducing room temperature, and increase thermal comfort level	[54]
Japan	Granules containing paraffin	PCM packed bed storage	Each night, 89% of daily cooling load is stored in 30-mm-thick packed bed of granular PCM	[55]
Japan (8 cities)	PCM granules (65% ceramic and 35% paraffinic hydrocarbon)	Vertically packed PCM bed fixed in a supply air duct	Reduction in ventilation load in all cities, significant reduction in Kyoto (about 62.8%)	[56]
Sweden	Commercially available salt based PCM	Bulk PCM tank with a finned pipe	75% of the cooling demands covered by half of the electricity consumption	[57]

absorption cooling were explored. Solar cooling systems are known to reduce the need for cooling in building units in regions with hot climate and achieve conditions comfortable in summer at low primary consumption of energy [58]. By this, they are decreasing the need for electricity which can help decrease pollution. The enhancements of solar cooling system with PCM enable cooling even when the solar energy is not present. Results of some studies are presented in Table 6.5.

Naganao [62] proposed an innovative floor integrated air-conditioning system that shares cooling load over the daytime with active air conditioner. The working principle of the system is shown in Fig. 6.16. A commercially available microencapsulated PCM was used and packed under the floor. The room temperature was at 28 °C. The heat was stored during the night in microencapsulated PCM under the floor and next day reduced cooling load up to 92%.

In addition to thermo-physical properties of PCMs and encapsulation thickness, other factors that affect the efficiency of the free cooling system are inlet and outlet temperature of storage module. The temperature of the air that leaves PCM unit should be in thermal comfort range, 20–27 °C; therefore, recommended melting

Table 6.5 Research bases on climate condition for solar cooling

Location	PCM solar system	Results	References
European climatic conditions	PCM with dry re-cooled sorption chiller—LHS module with inner heat exchanger with 1 m^3 of PCM	Positive effect on seasonal energy efficiency ratio for cooling by 11.4 (in situ measurement), efficiency up to 64% compared to the system without PCM (simulation result)	[59]
Lleida and Seville, Spain	PCM storage tank with absorption chiller and fresnel collectors, hydroquinone as PCM	Tank with fins: shorter melting/solidification time, energy stored faster Fins: money and time investment, rejection of using fins in real applications	[60]
52 provinces of Spain	PCM in the heat rejection loops of absorption chillers	Alternative system with TES_{pcm} of 1 m^3 can enhance the system performance coefficient (for locations with humid summers by almost one unit), reduction of total cooling energy in the evaporator (21–38%)	[61]

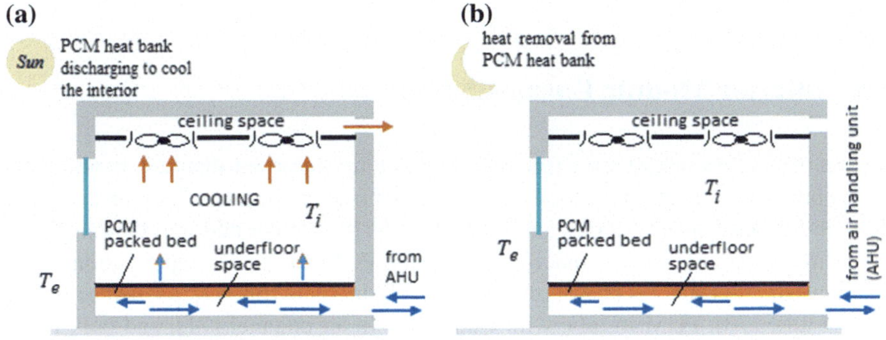

Fig. 6.16 PCM packed bed in floor for free cooling

temperature for free cooling system is from 19 to 24 °C. Published research revealed that commercially available PCMs between 20 and 27 °C were most frequently used in passive building heating/cooling systems. Among these, paraffins are the most used PCMs, since they do not react with encapsulation and as that the system is safe from leakage. Free cooling systems are technically and economically beneficial with

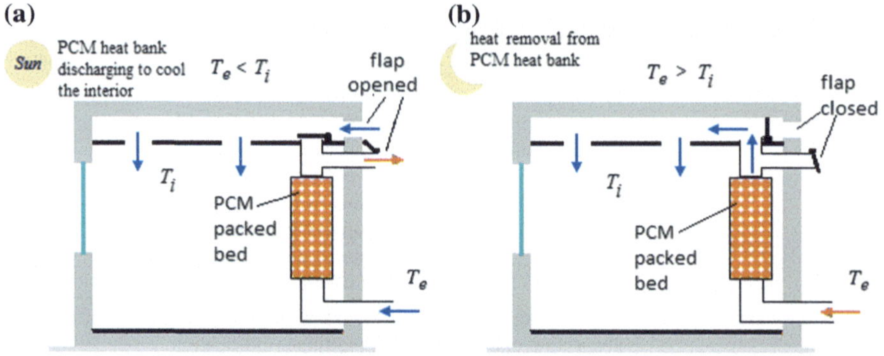

Fig. 6.17 PCM packed bed in building ventilation system

a potential to increase the efficiency by applying some of the heat transfer techniques [57].

Packed bed modules are applicable for building ventilation systems and have the capacity to reduce ventilation load up to 62% [63]. The integration of RT20 paraffin capsules in building ventilation system is investigated by Arkar and Medved [64]. Graphical representation and system operation over a day cycle are shown in Fig. 6.17.

The authors kept phase change temperature within the operating range of ±2 °C of the operating temperature. To ensure maximum cooling degree hours, they concluded that optimal mass of PCM is about 6.4 kg/m^2 of floor area with specific air follow rate of 0.7 m^3/h kg of PCM.

6.8 Interior Module Encapsulation

Design of heat exchangers for free cooling is more important than enhancing thermal conductivity of PCMs [65]. The potential of free cooling has been investigated through various capsule shapes and arrangements [66] integrated into building interior. Different geometries of macroencapsulated PCM for interior applications have been investigated such as rectangular tubes [67], flat plate [68, 69], packed bed [65], tube in tube [70], PCM in tube with water flow in air channel [71], and bulk PCM tank with finned heat exchanger [72]. Air is used as a HTF for interior applications. Some of these basic solutions are shown in Figs. 6.18 and 6.19.

Figures 6.18 and 6.19 show several solutions for encapsulation of PCM in interior. In each case, forced ventilation is used to enhance convection between PCM module and surrounding air. Heat transfer enhancement techniques may be applied to this basic solution that includes longitudinal or ring fins, nanoparticles, or other enhancements. This will reduce the amount of the PCM applied in the module and increase heat transfer between the module and the air.

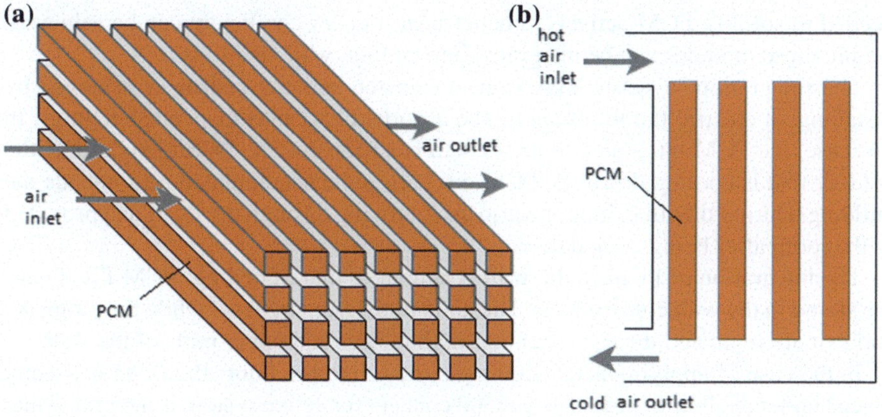

Fig. 6.18 PCM capsulation used for interior applications: **a** rectangular tubes filled with PCM; **b** tube in tube; **c** flat plate exchanger

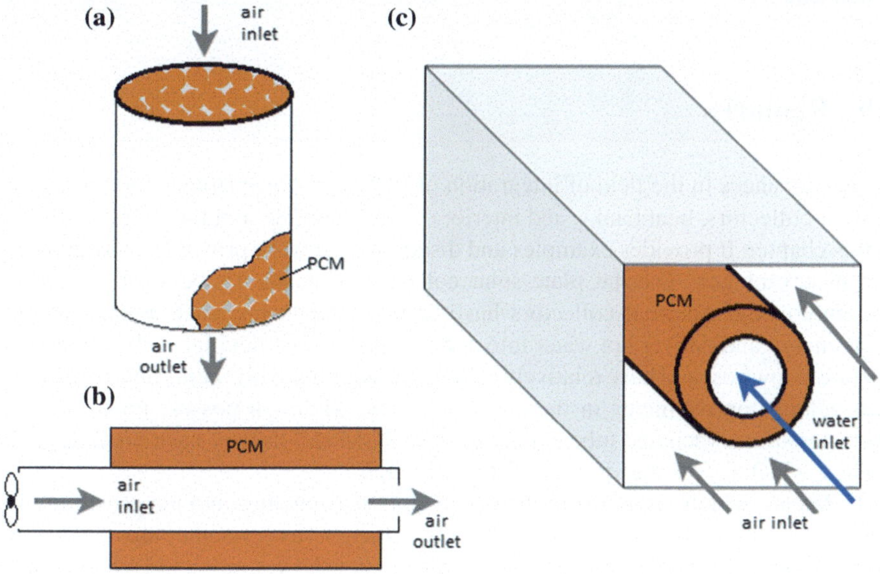

Fig. 6.19 PCM capsulation used for interior applications: **a** packed bed; **b** tube in tube; **c** PCM tube in channel with water flow

6.8.1 PCM Active System Applications in Building Envelope

Active PCM-based systems are developed by the integration of mechanical equipment and turning convection from its free to forced mode. With the minimum energy needed to solidify PCM actively, a better heat transfer coefficient can be achieved. Again, these modules can be integrated into ceiling, wall, and floor.

The integration of active system heat transfer into the **ceiling** is achieved by installing an electric fan to discharge the absorbed heat into suspended ceilings. In this case, the PCM integrated in the system is a heat storage. During nighttime, the cold air that is cooling down the PCM discharges the accumulated heat outside the building. During daytime, indoor warm air is moving through the PCM and provided to the room after being cooled down.

The integration of PCM in the **wall** is similar to that of ceiling. PCM-filled bags are stored in the wall construction. The air is transferred by fans, while the air intake and exit are regulated through openings at the lower and upper parts of the wall.

In the case of applying active cooling system into the **floor**, the PCM is usually placed under the floorboards. The general concept for active system is integrated into the floor.

Similar to other concepts of integrating active system, during nighttime when cold air is circulating under the floorboards, it cools down the PCM and discharges the stored heat.

6.9 Remarks

Recent advances in the field of integration of PCMs into heat storage modules such as solar collectors, heat banks, and interior devices have been in focus of discussion in this chapter. It provides examples and discussion on work principles, advantages, and disadvantages. The flat plate solar collectors evacuated tube solar collectors and photovoltaic/thermal collectors have certain advantages, such as they utilize solar energy and deliver hot water into needed spaces. The flat plate solar collectors operate at low cost and have relatively easy manufacturing while being able to achieve satisfactory improvements in thermal conditions. Although possess great thermal characteristics, evacuated tube collectors with PCM still did not have breakthrough studies, which leaves the space for future research.

PCMs are widely used due to their high storage capacities and the possibility to maintain a constant temperature during a phase change. However, they also have a low thermal conductivity that causes low heat transfer, low heat storage, and release rates. This can be improved by using different enhancement techniques. Studies mentioned in this chapter conclude that the encapsulation of PCMs results with high thermal conductivity prolongs usage and solves leakage. Furthermore, the results indicated that the thermal efficiency of heat banks can be improved by adding high thermal conductivity materials such as metallic foams and expanded graphite. However, the addition of different elements can also have negative impacts, as for example, metal-based materials can lead to poor stability. The different methods of enhancing PCMs can also have disadvantages that leave the space for future exploration and further enhancements of PCMs.

References

1. Durakovic B, Torlak M (2017) Simulation and experimental validation of phase change material and water used as heat storage medium in window applications. J Mater Environ Sci 8(5):1837–1846
2. Mehling H, Cabeza LF (2008) Heat and cold storage with PCM: an up to date introduction into basics and applications. Springer, Berlin
3. BINE Information Service. Concepts for building services technology. [Online]. Available: http://www.bine.info/en/publications/publikation/latentwaermespeicher-in-gebaeuden/pcm-konzepte-fuer-die-gebaeudetechnik/. Accessed 16 Feb 2014
4. Farid MM, Khudhair AM, Razack SAK, Al-Hallaj S (2004) A review on phase change energy storage: materials and applications. Energy Convers Manag 45(9–10):1597–1615
5. Kissock J, Kelly J, Hannig M, Thomas I (1998) Testing and simulation of phase change wallboard for thermal storage in buildings. In: Proceedings of international solar energy conference, Albuquerque, New Mexico, 14–17 June 1998
6. Khan MMA, Ibrahim NI, Mahbubul I, Ali HM, Saidur R, Al-Sulaiman FA (2018) Evaluation of solar collector designs with integrated latent heat thermal energy storage: a review. Sol Energy 166:334–350
7. Durakovic B, Torlak M (2017) Experimental and numerical study of a PCM window model as a thermal energy storage unit. Int J Low-Carbon Technol 12(3):272–280
8. Durakovic B, Yildiz G, Yahia ME (2020) Comparative performance evaluation of conventional and renewable thermal insulation materials used in building envelops. Tehnicki vjesnik - Tech Gaz 27(1) (in Press)
9. Bouadila S, Fteïti M, Oueslati MM, Guizani A, Farhat A (2014) Enhancement of latent heat storage in a rectangular cavity: solar water heater case study. Energy Convers Manag 78:904–912
10. Lin S, Al-Kayiem H, Bin Aris M (2012) Experimental investigation on the performance enhancement of integrated PCM-flat plate solar collector. J Appl Sci 12(23):2390–2396
11. Lee WS, Chen BR, Chen SL (2006) Latent heat storage in a two-phase thermosyphon solar water heater. J Solar Energy Eng 128(1):69–76
12. Khalifa AJN, Suffer KH, Mahmoud MS (2013) A storage domestic solar hot water system with a back layer of phase change material. Exp Therm Fluid Sci 44:174–181
13. Serale G, Baronetto S, Goia F, Perino M (2014) Characterization and energy performance of a slurry PCM-based solar thermal collector: a numerical analysis. In: International conference on solar heating and cooling for buildings and industry, Freiburg, Germany
14. Gao Y, Zhang Q, Fan R, Lin X, Yu Y (2013) Effects of thermal mass and flow rate on forced-circulation solar hot-water system: comparison of water-in-glass and U-pipe evacuated-tube solar collectors. Solar Energy 98:290–301
15. Riffat S, Jiang L, Zhu J, Gan G (2006) Experimental investigation of energy storage for an evacuated solar collector. Int J Low-Carbon Technol 1(2):139–148
16. Naghavi M, Ong KS, Badruddin IA, Mehrali M, Silakhori M, Metselaar HSC (2015) Theoretical model of an evacuated tube heat pipe solar collector integrated with phase change material. Energy 91:911–924
17. Mehla N, Yadav A (2015) Experimental analysis of thermal performance of evacuated tube solar air collector with phase change material for sunshine and off-sunshine hours. Int J Ambient Energy 38(2):130–145
18. Papadimitratos A, Sobhansarbandi S, Pozdina V, Zakhidovc A, Hassanipour F (2016) Evacuated tube solar collectors integrated with phase change materials. Sol Energy 129:10–19
19. Feliński P, Sekret R (2016) Experimental study of evacuated tube collector/storage system containing paraffin as a PCM. Energy 114:1063–1072
20. Makki A, Omer S, Sabir H (2015) Advancements in hybrid photovoltaic systems for enhanced solar cells performance. Renew Sustain Energy Rev 41:658–684

21. Hasan A, McCormack S, Huang M, Norton B (2010) Evaluation of phase change materials for thermal regulation enhancement of building integrated photovoltaics. Sol Energy 84(9):1601–1612
22. Huang M, Eames P, Norton B (2004) Thermal regulation of building-integrated photovoltaics using phase change materials. Int J Heat Mass Transf 47(12–13):2715–2733
23. Malvi CS, Dixon-Hardy DW, Crook R (2011) Energy balance model of combined photovoltaic solar-thermal system incorporating phase change material. Sol Energy 85(7):1440–1446
24. Su D, Jia Y, Alva G, Liu L, Fang G (2017) Comparative analyses on dynamic performances of photovoltaic–thermal solar collectors integrated with phase change materials. Energy Convers Manag 131(1):79–89
25. Hasan A, McCormack S, Huang M, Sarwar J, Norton B (2015) Increased photovoltaic performance through temperature regulation by phase change materials: materials comparison in different climates. Sol Energy 115:264–276
26. Regin AF, Solanki S, Saini J (2008) Heat transfer characteristics of thermal energy storage system using PCM capsules: a review. Renew Sustain Energy Rev 12(9):2438–2458
27. Zayed ME, Zhao J, Elsheikh AH, Hammad FA, Ma L, Du Y, Kabeel A, Shalaby S (2019) Applications of cascaded phase change materials in solar water collector storage tanks: a review. Solar Energy Mater Solar Cells 199:24–49
28. Mahfuz M, Anisur M, Kibria M, Saidur R, Metselaar I (2014) Performance investigation of thermal energy storage system with phase change material (PCM) for solar water heating application. Int Commun Heat Mass Transf 57:132–139
29. Shirinbakhsh M, Mirkhani N, Sajadi B (2018) A comprehensive study on the effect of hot water demand and PCM integration on the performance of SDHW system. Solar Energy 159:405–414
30. Agyenim F, Hewitt N, Eames P, Smyth M (2010) A review of materials, heat transfer and phase change problem formulation for latent heat thermal energy storage systems (LHTESS). Renew Sustain Energy Rev 14(2):615–628
31. Shukla R, Sumathy K, Erickson P, Gong J (2013) Recent advances in the solar water heating systems: a review. Renew Sustain Energy Rev 19:173–190
32. Horbaniuca B, Dumitrascu G, Popescu A (1999) Mathematical models for the study of solidification within a longitudinally finned heat pipe latent heat thermal storage system. Energy Convers Manag 40(15–16):1765–1774
33. Agyenim F, Eames P, Mervyn S (2009) A comparison of heat transfer enhancement in a medium temperature thermal energy storage heat exchanger using fins. Sol Energy 83(9):1509–1520
34. Papanicolaou E, Belessiontis V (2001) Transient natural convection in a cylindrical enclosure at high Rayleigh numbers. Heat Mass Transf 45(7):1425–1444
35. Silva PD, Gonçalves L, Pires L (2002) Transient behaviour of a latent-heat thermal energy store: numerical and experimental studies. Appl Energy 73(1):83–98
36. Zivkovic B, Fujii I (2001) An analysis of isothermal phase change material within rectangular and cylindrical containers. Sol Energy 70(1):51–56
37. Lin Y, Alva G, Fang G (2018) Review on thermal performances and applications of thermal energy storage systems with inorganic phase change materials. Energy 165(Part A):685–708
38. Wu M, Xu C, He Y-L (2014) Dynamic thermal performance analysis of a molten-salt packed-bed thermal energy storage system using PCM capsules. Appl Energy 121:184–195
39. Muñoz-Sánchez B, Iparraguirre-Torres I, Madina-Arrese V, Izagirre-Etxeberria U, Unzurrunzaga-Iturbe A, García-Romero A (2015) Encapsulated high temperature PCM as active filler material in a thermocline-based thermal storage system. Energy Procedia 69:937–946
40. Cardenas B, Leon N (2013) High temperature latent heat thermal energy storage: phase change materials, design considerations and performance enhancement techniques. Renew Sustain Energy Rev 27:724–737
41. Qureshi ZA, Ali HM, Khushnood S (2018) Recent advances on thermal conductivity enhancement of phase change materials for energy storage system: a review. Int J Heat Mass Transf 127(Part C):838–856

42. Sari A, Karaipekli A (2007) Thermal conductivity and latent heat thermal energy storage characteristics of paraffin/expanded graphite composite as phase change material. Appl Therm Eng 27(8–9):1271–1277
43. Ceng J, Sun L, Xu F, Tan Z, Zhang T (2005) Study of a PCM based storage system containing metal nanoparticles. In: 4th international and 6th Japan-China joint symposium on calorimetry and thermal analysis, Japan
44. Shin D, Banerjee D (2009) Investigation of nanofluids for solar thermal storage applications. In: Proceedings of the ASME 2009 3rd international conference of energy sustainability, San Francisco, California, USA, pp 819–822
45. Saw C, Al-Kayiem H, Owolabi A (2013) Experimental investigation on the effect of PCM and nano-enhanced PCM of integrated solar collector performance. In: WIT transactions on ecology and the environment. WIT Press, UK, pp 899–909
46. Ma Z, Lin W, Sohel M (2016) Nano-enhanced phase change materials for improved building performance. Renew Sustain Energy Rev 58:1256–1268
47. Dincer I, Rosen M (2010) Thermal energy storage: systems and applications. 2nd edn. Wiley, West Sussex
48. Sharma A, Tyagi VV, Chen CR, Buddhi D (2009) Review on thermal energy storage with phase change materials and applications. Renew Sustain Energy Rev 13(2):318–345
49. Souayfane F, Fardoun F, Biwole P-H (2016) Phase change materials (PCM) for cooling applications in buildings: a review. Energy Build 129:396–431
50. Zalba B, Marín JM, Cabeza LF, Mehling H (2004) Free-cooling of buildings with phase change materials. Int J Refrig 27(8):839–849
51. Waqas A, Din ZU (2013) Phase change material (PCM) storage for free cooling of buildings—a review. Renew Sustain Energy Rev 18:607–625
52. Lee K, Medina M, Raith E, Sun X (2015) Assessing the integration of a thin phase change material (PCM) layer in a residential building wall for heat transfer reduction and management. Appl Energy 137:699–706
53. Osterman E, Butala V, Stritih U (2015) PCM thermal storage system for 'free' heating and cooling of buildings. Energy Build 106:125–133
54. Yanbing K, Yi J, Yinping Z (2003) Modeling and experimental study on an innovative passive cooling system—NVP system. Energy Build 35(4):417–425
55. Nagano K, Takeda S, Mochida T, Shimakura K, Nakamura T (2006) Study of a floor supply air conditioning system using granular phase change material to augment building mass thermal storage—heat response in small scale experiments. Energy Build 38(5):436–446
56. Takeda S, Nagano K, Mochida T, Shimakura K (2004) Development of a ventilation system utilizing thermal energy storage for granules containing phase change material. Sol Energy 77(3):329–338
57. Chiu JNW, Gravoille P, Martin V (2013) Active free cooling optimization with thermal energy storage in Stockholm. Appl Energy 109:523–529
58. Henning H-M (2007) Solar assisted air conditioning of buildings—an overview. Appl Therm Eng 27(10):1734–1749
59. Helm M, Hagel K, Pfeffer W, Hiebler S, Schweigler C (2014) Solar heating and cooling system with absorption chiller and latent heat storage—a research project summary. Energy Procedia 48:837–849
60. Gil A, Oró E, Miró L, Peiró G, Ruiz Á, Salmerón JM, Cabeza LF (2014) Experimental analysis of hydroquinone used as phase change material (PCM) to be applied in solar cooling refrigeration. Int J Refrig 39:95–103
61. Belmonte J, Izquierdo-Barrientos M, Eguía P, Molina A, Almendros-Ibáñez J (2014) PCM in the heat rejection loops of absorption chillers. A feasibility study for the residential sector in Spain. Energy Build 80:331–351
62. Naganao K (2007) Development of the PCM floor supply air-conditioning system. In: Paksoy HÖ (ed) Thermal energy storage for sustainable energy consumption, vol 234. NATO science series (mathematics, physics and chemistry). Springer, Dordrecht, pp 367–373

63. Takeda S, Nagano K, Mochida T, Shimakura K (2004) Development of a ventilation system utilizing thermal energy storage for granules containing phase change material. Solar Energy 77:329–338
64. Arkar C, Medved S (2007) Free cooling of a building using PCM heat storage integrated into the ventilation system. Solar Energy 81:1078–1087
65. Lazaro A, Dolado P, Marín J, Zalba B (2009) PCM-air heat exchangers for free-cooling applications in buildings: experimental results of two real-scale prototypes. Energy Convers Manag 50(3):439–443
66. Torlak M, Delalić N, Durakovic B, Gavranović H (2014) CFD-based assessment of thermal energy storage in phase-change materials—(PCM). In: International energy technologies conference proceedings 2014—ENTECH'2014, Istanbul, Turkey
67. Rouault F, Bruneau D, Sebastian P, Lopez J (2013) Numerical modelling of tube bundle thermal energy storage for free-cooling of buildings. Appl Energy 111:1099–1106
68. Duraković B, Mešetović S (2019) Thermal performances of glazed energy storage systems with various storage materials: an experimental study. Sustain Cities Soc 45:422–430
69. Darzi CA, Moosania S, Tan F, Farhadi M (2013) Numerical investigation of free-cooling system using plate type PCM storage. Int Commun Heat Mass Transf 48:155–163
70. Anisur CM, Kibria M, Mahfuz M, Saidur R, Metselaar I (2014) Cooling of air using heptadecane phase change material in shell and tube arrangement: analytical and experimental study. Energy Build 85:98–106
71. Tan G, Zhao D (2015) Study of a thermoelectric space cooling system integrated with phase change material. Appl Therm Eng 86:187–198
72. Chiu J, Gravoille P, Martin V (2013) Active free cooling optimization with thermal energy storage in Stockholm. Appl Energy 109:523–529

Chapter 7
Heat Transfer Mechanisms in PCM-Based Building Envelope Systems

Nomenclature

C	Coefficient
g	Gravity acceleration, m/s^2
U	Overall heat transfer coefficient/U-factor, W/m^2 K
H	Height, m
A	Area, m^2
c_p	Specific heat, J/kg °C
T	Temperature, °C
h	Convective heat transfer coefficient, W/m^2 K
β	Fraction of melted PCM
m	Mass, kg
$\underset{\cdot}{\dot{q}}$	Heat flux in positive direction, W/m^2
$\dot{Q}$	Heat flow, W
$\dot{q}$	Heat flux, W/m^2
I	Solar irradiance, W/m^2
E	Energy, J
σ	Stefan–Boltzmann constant ($\sigma = 5.67 \times 10^{-8}$ W/m^2 K^4)
ε	Surface emissivity
λ	Thermal conductivity of material, W/m K
$\dot{E}$	Rate of heat, W
$\dot{e}$	Rate of heat, W/m^3
α	Thermal diffusivity, m^2/s
ρ	Density, kg/m^3
L	Thickness/length, m
u	Velocity in x-direction, m/s
v	Velocity in y-direction, m/s
w	Velocity in z-direction, m/s
F	Force
τ	Shear stress tensor, Pa

© Springer Nature Switzerland AG 2020

B. Duraković, *PCM-Based Building Envelope Systems*, Green Energy and Technology, https://doi.org/10.1007/978-3-030-38335-0_7

$\hat{\tau}$	Transmissivity of the surface
t	Time, s
μ	Dynamic viscosity of the liquid phase, Pa s
ψ_j	Linear heat transfer coefficient, W/m K
χ_k	Dotted heat transfer coefficient, W/m K
p	Number of dotted connections.

Abbreviations

PCM	Phase change material
LHES	Latent heat energy storage
SHES	Sensible heat energy storage.

Subscripts

i	Interior
e	Exterior
s	Surface
surr	Surrounding
in	Inlet
out	Outlet
is	Interior side
d	Discharge
sc	Solar chimney
pc	Phase change
cond	Conduction
conv	Convection
R	Radiation
c	Convection
T	Total
cv	Control volume
gen	Generated
stored	Stored
tot	Total.

Heat is a form of energy that can be transferred from one system to another due to temperature differences. The heat flow takes place from the higher-temperature body to lower-temperature body. The process of transferring the heat stops once the body reaches the temperature equilibrium. The energy at the molecular level of activity is called internal energy and can be in one of two forms: sensible or latent. Sensible energy is associated with the kinetic energy of the particles and their degree of activities. Latent energy is the energy absorbed or released from a body due to its phase change [1, 2]. If sufficient energy is added to a solid body, the intermolecular forces become weak turning the body to liquid (or liquid system to gas as well). This transformation does not cause a change in the chemical composition of a body.

The amount of energy stored as sensible heat in a body and changing the body temperature from T_1 to T_2 can be expressed as [3]:

$$Q = \int_{T_1}^{T_2} mC_p \mathrm{d}T = mC_p(T_2 - T_1) \tag{7.1}$$

The amount of energy absorbed as latent heat during phase transition of PCM can be expressed mathematically as:

$$Q = m\beta_{pc}\Delta h_{pc} \tag{7.2}$$

where m is mass in kg, C_p specific heat capacity of the body in J/kg K, T_1 and T_2 temperatures of the body before and after heating in K, Δh_{pc} enthalpy of phase change material in J/kg, and β_{pc} a melted PCM fraction.

Heat transfer rate is the amount of heat transfer per unit of time, and it is denoted by $\dot{Q}$. The dot over means per unit of time and has the unit J/s, which is equivalent to watt (W). Heat transfer rate per unit area is called heat flux and expressed as:

$$\dot{q} = \frac{\dot{Q}}{A} \text{ in } \left(\mathrm{W/m^2}\right) \tag{7.3}$$

where A is heat transfer area in m^2, and $\dot{q}$ is heat flux in positive direction (W/m^2).

The total amount of heat transfer during a time period Δt can be calculated as

$$Q = \int_{0}^{\Delta t} \dot{Q} \mathrm{d}t \text{ in (J)} \tag{7.4}$$

Heat transfer can be steady or transient (unsteady). In steady state, there is no heat flux change over time, while in transient state there is heat flux change over time through the wall. In practice, most of the heat transfer problems through the building envelope are transient in nature, but they are analyzed as steady-state problems for some extreme cases.

Heat transfers can appear in one of three modes: conduction, convection, and radiation, which is discussed in detail in the following pages.

7.1 Airflow Modeling Through Solar Chimneys

Various mathematical models for solar chimney were developed bases on numerical and experimental results and discussed in the literature. These models can be utilized for the solar chimney performance prediction. Working principle and volumetric flows through solar chimneys are shown in Table 7.1.

Nomenclature in the previous formula is defined as: V is the flow rate in m^3/s; A is the area in m^2; C_i is interior configuration coefficient; C_p is the specific heat capacity in J/kg °C; α is the thermal expansion coefficient in 1/K; g is the acceleration of gravity in m/s^2; k is the coefficient of pressure loss; f is the coefficient of wall friction; H is the solar chimney height in m; d is solar chimney gap thickness, m; T is the temperature in K; θ is the inclination angle; ρ is air density in kg/m^3; h is the convective heat transfer coefficient in W/m K. Subscripts: sc—solar chimney; d—discharge; in—inlet; out—outlet.

Table 7.1 Solar chimney mathematical models

Type	Mathematical model	References
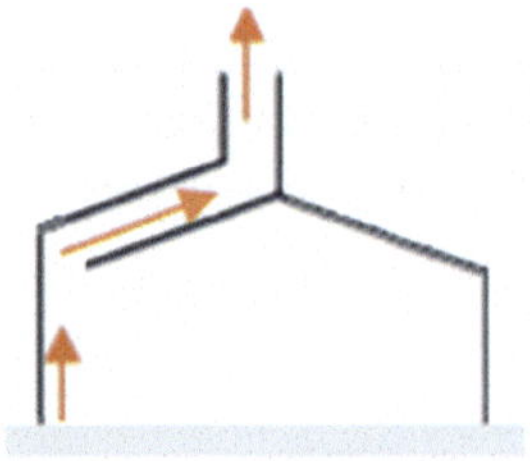	$V = \frac{A_{out}\sqrt{2\alpha g H(T_{sc}-T_{in})}}{\sqrt{k_{in}\left(\frac{A_{out}}{A_{in}}\right)^2 + k_{out} + f\frac{H}{d}}}$	[4]
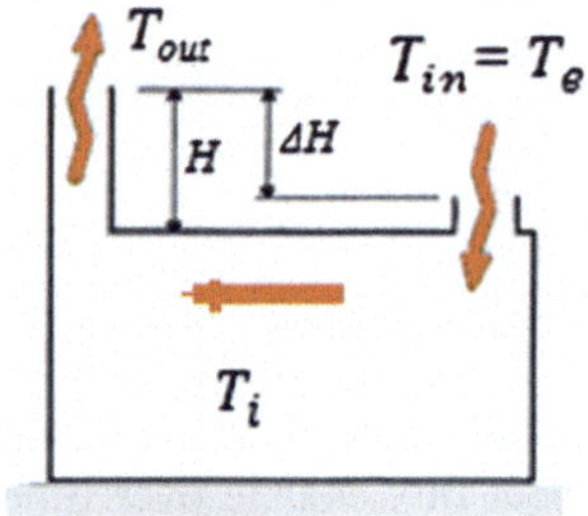	$V = C_d A\sqrt{\frac{gH(T_{out}-T_{in})}{T_{out}+T_{in}}}$	[5]
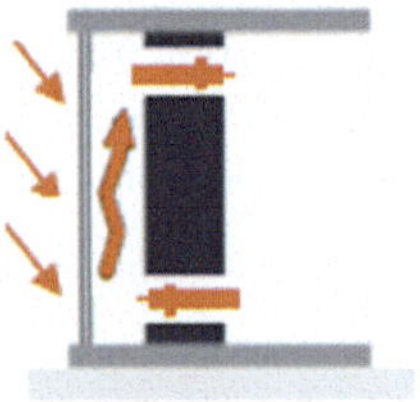	$V = C_d \sin\theta \frac{\rho_{sc}}{\rho_{in}} A_{in}\sqrt{\frac{gH(T_{sc}-T_{in})}{T_{in}}}$	[6]

(continued)

Table 7.1 (continued)

Type	Mathematical model	References
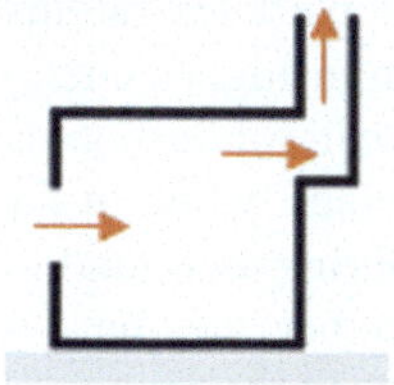	$V = 23.37 A_{\text{in}} A_{\text{out}} \sqrt{\frac{gH(T_{\text{sc}}-T_{\text{in}})}{(k_{\text{in}} A_{\text{out}}^2 + k_{\text{out}} A_{\text{in}}^2) T_{\text{sc}} T_{\text{in}}}}$	[7]
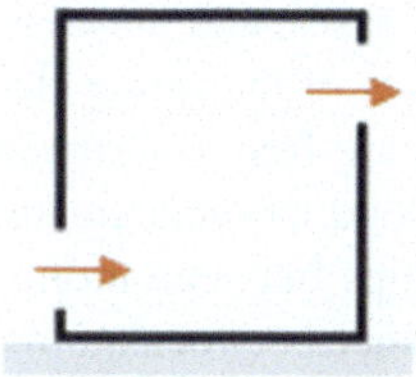	$V = C_{\text{d}} A_{\text{in}} \sqrt{\frac{2gH(T_{\text{out}}-T_{\text{in}})}{T_{\text{out}}}}$	[8]
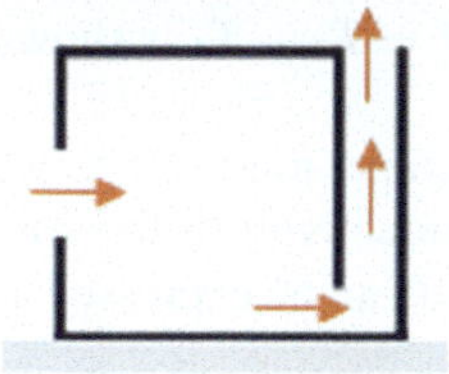	$V = C_i \sqrt[3]{hH(T_{\text{sc}} - T_{\text{in}})A_{\text{sc}}}$	[9]

Table 7.1 shows a summary of mathematical models that can be used for solar chimney performance evaluation. Generally, three groups of solar chimneys are represented: roof solar chimney, wall solar chimney, and combined solar chimneys. According to the literature, four groups of influencing factors are important for the performance of solar chimney: environment, materials used, configuration, and installation conditions.

The influence of environment depends on climate conditions such as solar radiation intensity; building orientation and internal heat absorption; external wind speed has a significant impact on ventilation rate through a solar chimney. Influence of materials used on the solar chimney performances depends on glazing type, the material of solar absorber and thermal mass (latent or sensible heat), insulation of thermal mass to prevent heat loss to exterior by thermal radiation and convection. The configuration depends on height, cavity gap, and inlet and outlet cross-sectional areas. Installation conditions depend on the inclination angle, which is the angle between solar chimney and the horizontal plane (e.g., for wall solar chimney, it is 90°). An optimum angle is one of the most important influencing factors on the performance. The other configuration factors are opening of the room which allows inlet of fresh air and different styles of solar collectors (vertical/horizontal/inclined).

7.2 Heat Conduction Through Building Envelope

Thermal conduction is the heat transfer by the interaction of particles (atoms, molecules, etc.) of a material. The heat is being transferred from higher energy particles to adjacent lover energy particles. Conduction takes place in solids, liquids, or gasses. Conduction in solids is due to the vibration of molecules, while in liquids and gasses it is due to collision and diffusion of molecules during their random motion. Transferring the heat by this mode from one body to another, they have to be in contact.

Conduction takes place in all forms of matter: solids, liquids, gases, and plasmas. In solids, heat conduction is due to the combination of vibrations of the molecules in a lattice and diffusion of free electrons. In fluids, molecules are free to move in space, and heat is transferred by direct molecular interactions. Conduction is greater in solids because of the relatively close fixed spatial relationships between atoms, whereas fluids and gases are less conductive due to the large distance between atoms.

Steady heat conduction rate $\dot{Q}_{\mathrm{cond}}$ through a wall of surface area A and thickness $\Delta x = L$ that has the temperature difference across the wall $\Delta T = T_2 - T_1$ is shown in Fig. 7.1b.

Heat conduction through the wall is proportional to the temperature difference across the wall ΔT, wall surface area A, average thermal conductivity of the wall material, and inversely proportional to the wall thickness Δx. It can be expressed as follows:

$$\dot{Q}_{\mathrm{cond}} = -\lambda A \frac{\Delta T}{\Delta x} = -\lambda A \frac{T_2 - T_1}{L} \tag{7.5}$$

where

$\dot{Q}_{\mathrm{cond}}$ heat conduction rate in positive direction, in W

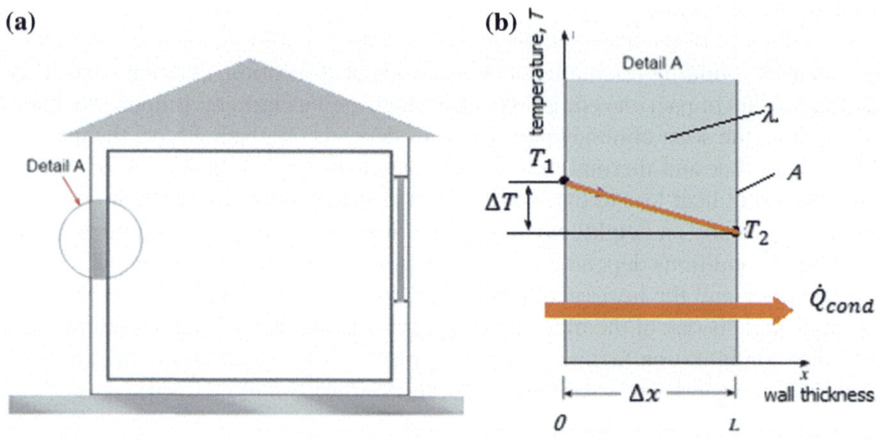

Fig. 7.1 Heat transfer through building envelope

λ thermal conductivity of the wall material, W/m K
A wall surface area, m^2.

Negative sign in Eq. (7.5) ensures that heat transfer in positive x-direction has a positive quantity.

7.2.1 Fourier Law of Heat Conduction

Fourier developed the fundamental laws of heat conduction from experimental observations on steady-state systems. Assuming that there is no heat generation in the wall, steady-state energy balance equation can be written as

$$\dot{Q}_{\text{in}} = \dot{Q}_{\text{out}} \tag{7.6}$$

This means that the rate of heat transfer that enters the wall is equal to the rate of heat transfer that leaves the wall.

The temperature change from the exterior surface to the interior surface (from T_1 to T_2) usually has nonlinear characteristics. Thus, the differential form of the heat transfer is a good approximation of the temperature characteristic (Fig. 7.2).

Temperature gradient $\frac{\mathrm{d}T}{\mathrm{d}x}$ (K/m), at any point across the wall thickness, is given by the slope of the tangent at that point, which is the rate of change of temperature T with length x in T–x diagram (Fig. 7.2).

$$\dot{Q}_{\text{cond}} = -\lambda A \frac{\mathrm{d}T}{\mathrm{d}x} \tag{7.7}$$

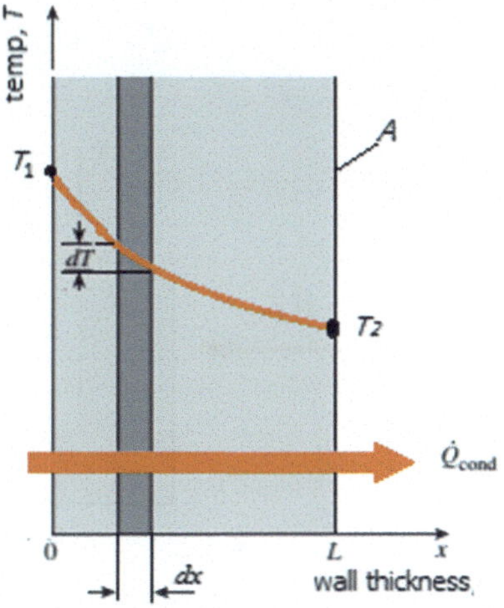

Fig. 7.2 Temperature gradient

Multiplying Eq. (7.7) by dx and then integrating from $x = 0$ to $x = L$, and from $T(0) = T_1$ to $T(L) = T_2$, the following is obtained

$$\int_{x=0}^{L} \dot{Q}_{\text{cond}} \mathrm{d}x = -\int_{T=T_1}^{T_2} \lambda A \mathrm{d}T \quad (7.8)$$

After integration of Eq. (7.8), and dividing by L

$$\dot{Q}_{\text{cond}} = \lambda A \frac{(T_1 - T_2)}{L} \quad (\mathrm{W}) \quad (7.9)$$

Heat flux

$$\dot{q} = -\lambda \frac{\mathrm{d}T}{\mathrm{d}x} \quad (7.10)$$

Temperature gradient has negative value when temperature decreases with an increase of x, and heat flow takes place in the direction of decreasing temperature. The same analogy is for heat fluxes in positive x-, y-, z-directions.

$$\dot{q}_x = -\lambda \frac{\partial T}{\partial x}; \quad \dot{q}_y = -\lambda \frac{\partial T}{\partial y}; \quad \dot{q}_z = -\lambda \frac{\partial T}{\partial z} \quad (7.11)$$

Energy conservation concept is applied to very small region of elemental control volume (dx, dy, dz) (Fig. 7.3).

Energy balance equation for the control volume during a very small time interval Δt is

$$\dot{Q}_{\text{in}} - \dot{Q}_{\text{out}} + \dot{E}_{\text{gen,cv}} = \frac{\Delta \dot{E}_{\text{stored,cv}}}{\Delta t}$$

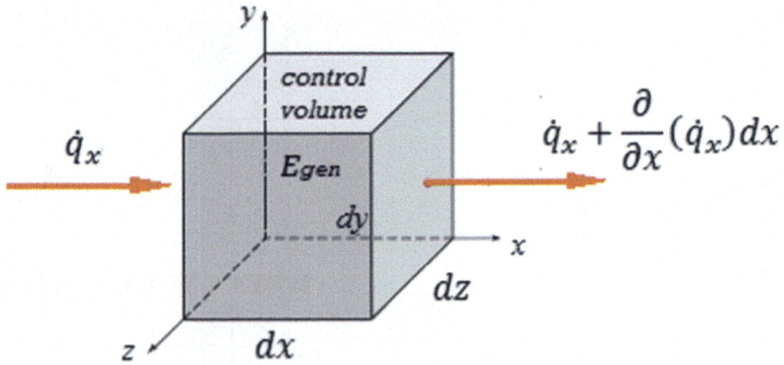

Fig. 7.3 A small control volume in Cartesian coordinate system

where $\dot{Q}_{\mathrm{in}}$ is heat transfer rate that enter control volume in W, $\dot{Q}_{\mathrm{out}}$ is heat transfer rate that leaves the control volume, $\dot{E}_{\mathrm{gen,cv}}$ is rate of heat generated inside the control volume, $\Delta\dot{E}_{\mathrm{stored,cv}}$ is rate of heat change in the control volume, and Δt is a small time interval during which the heat change takes place.

One-dimensional energy balance equations in a differential form through each side of the control volume are defined as follows:

Heat flow in x-direction

Heat passing through the $\mathrm{d}y \cdot \mathrm{d}z$ surface of the control volume is

$$\dot{q}_x \mathrm{d}y \cdot \mathrm{d}z - \left[\dot{q}_x + \frac{\partial}{\partial x}(\dot{q}_x)\mathrm{d}x\right]\mathrm{d}y \cdot \mathrm{d}z + \dot{e}_{\mathrm{gen}}\, \mathrm{d}x \cdot \mathrm{d}y \cdot \mathrm{d}z = \rho c_{\mathrm{p}} \frac{\partial T}{\partial t} \mathrm{d}x \cdot \mathrm{d}y \cdot \mathrm{d}z \tag{7.12}$$

Heat flow in y-direction

Heat passing through the $\mathrm{d}x \cdot \mathrm{d}z$ surface of the control volume is

$$\dot{q}_y \mathrm{d}x \cdot \mathrm{d}z - \left[\dot{q}_y + \frac{\partial}{\partial y}(\dot{q}_y)\mathrm{d}y\right]\mathrm{d}x \cdot \mathrm{d}z + \dot{e}_{\mathrm{gen}}\, \mathrm{d}x \cdot \mathrm{d}y \cdot \mathrm{d}z = \rho c_{\mathrm{p}} \frac{\partial T}{\partial t} \mathrm{d}y \cdot \mathrm{d}x \cdot \mathrm{d}z \tag{7.13}$$

Heat flow in z-direction

Heat passing through the $\mathrm{d}x \cdot \mathrm{d}y$ surface of the control volume is

$$\dot{q}_z \mathrm{d}x \cdot \mathrm{d}y - \left[\dot{q}_z + \frac{\partial}{\partial z}(\dot{q}_z)\mathrm{d}z\right]\mathrm{d}x \cdot \mathrm{d}y + \dot{e}_{\mathrm{gen}}\, \mathrm{d}x \cdot \mathrm{d}y \cdot \mathrm{d}z = \rho c_{\mathrm{p}} \frac{\partial T}{\partial t} \mathrm{d}z \cdot \mathrm{d}x \cdot \mathrm{d}y \tag{7.14}$$

Adding (7.12), (7.13), and (7.14), and dividing by $\mathrm{d}x \cdot \mathrm{d}y \cdot \mathrm{d}z$, a **general heat conduction equation** in three dimensions is expressed as

$$-\left[\frac{\partial}{\partial x}(\dot{q}_x) + \frac{\partial}{\partial y}(\dot{q}_y) + \frac{\partial}{\partial z}(\dot{q}_z)\right] + \dot{e}_{\mathrm{gen}} = \rho c_{\mathrm{p}} \frac{\partial T}{\partial t} \tag{7.15}$$

where $\dot{e}_{\mathrm{gen}}$ is generated energy inside the element such as electrical resistance elements, heating coils in an electric oven, absorption solar radiation by water, chemical reaction, nuclear reaction. Heat generation is a volumetric phenomenon and expressed per unit volume (W/m^3). ρ is material density in kg/m^3. c_{p} is the specific heat of the material in J/kg K. Specific heat is defined as *energy required to raise the temperature of a unit mass of a substance by one degree*. The product ρc_{p} is called material heat capacity in J/m^3 K and represents the capability of the material to store the heat per unit of volume.

By replacing $\dot{q}$ in Eq. (7.15) with Eq. (7.10), the following form is obtained

$$\left[\frac{\partial}{\partial x}\left(\lambda\frac{\partial T}{\partial x}\right) + \frac{\partial}{\partial y}\left(\lambda\frac{\partial T}{\partial y}\right) + \frac{\partial}{\partial z}\left(\lambda\frac{\partial T}{\partial z}\right)\right] + \dot{e}_{\mathrm{gen}} = \rho c_{\mathrm{p}} \frac{\partial T}{\partial t} \tag{7.16}$$

or

$$\underbrace{\lambda\frac{\partial^2 T}{\partial x^2}+\lambda\frac{\partial^2 T}{\partial y^2}+\lambda\frac{\partial^2 T}{\partial z^2}}_{\substack{\text{Conduction in}\\ x,\,y,\,z\text{ directions}}}+\underbrace{\dot{e}_{\text{gen}}}_{\substack{\text{Internal}\\ \text{heat}\\ \text{generation}}}=\underbrace{\rho c_{\text{p}}\frac{\partial T}{\partial t}}_{\substack{\text{Thermal}\\ \text{inertia}}} \tag{7.17}$$

If it is assumed that material properties are independent of temperature, than material conductivity can be taken outside the derivative:

$$\lambda\left[\frac{\partial^2 T}{\partial x^2}+\frac{\partial^2 T}{\partial y^2}+\frac{\partial^2 T}{\partial z^2}\right]+\dot{e}_{\text{gen}}=\rho c_{\text{p}}\frac{\partial T}{\partial t} \tag{7.18}$$

or

$$\frac{\partial^2 T}{\partial x^2}+\frac{\partial^2 T}{\partial y^2}+\frac{\partial^2 T}{\partial z^2}+\frac{\dot{e}_{\text{gen}}}{\lambda}=\frac{\rho c_{\text{p}}}{\lambda}\frac{\partial T}{\partial t} \tag{7.19}$$

where $\alpha=\frac{\lambda}{\rho c_{\text{p}}}$ is **thermal diffusivity** of a material and represents how fast heat diffuses through the material. Materials with high thermal conductivity and low heat capacity will have a **high thermal diffusivity**, which leads to faster diffusion of the heat through the body. The **low value of thermal diffusivity** means that heat is mainly absorbed by a material as a small portion of the heat is conducted through the body.

Transient heat transfer—no heat generation (diffusion equation)

By plugging thermal diffusivity in (7.19) and assuming that there is no heat generation inside the wall $\left(\dot{e}_{\text{gen}}=0\right)$, a new form of transient heat transfer equation is obtained

$$\frac{\partial^2 T}{\partial x^2}+\frac{\partial^2 T}{\partial y^2}+\frac{\partial^2 T}{\partial z^2}=\frac{1}{\alpha}\frac{\partial T}{\partial t} \tag{7.20}$$

Steady-state heat conduction (Poisson and Laplace equations)

For steady-state conduction process, the time in the denominator tends to infinity $(t\to\infty)$, and thus, transient components of the heat transfer equation tend to be zero $\left(\rho c_{\text{p}}\frac{\partial T}{\partial t}\to 0\right)$. Therefore, Eq. (7.17) becomes

$$\frac{\partial^2 T}{\partial x^2}+\frac{\partial^2 T}{\partial y^2}+\frac{\partial^2 T}{\partial z^2}+\frac{\dot{e}_{\text{gen}}}{\lambda}=0 \tag{7.21}$$

Equation (7.21) is called Poisson equation.

For building physics purposes, internal heat generation component is usually neglected $\left(\dot{e}_{\text{gen}}=0\right)$, and Eq. (7.21) becomes

$$\frac{\partial^2 T}{\partial x^2} + \frac{\partial^2 T}{\partial y^2} + \frac{\partial^2 T}{\partial z^2} = 0 \tag{7.22}$$

Equation (7.22) is called Laplace equation and describes steady state of heat conduction through a wall with no heat generation.

Three-dimensional heat transfer equations above can be reduced to one dimensional in cases when the temperature changes in one dimension only.

7.2.2 *Thermal Conductivity*

Thermal conductivity λ is a measure of the ability of a material to conduct heat, and it is measured in W/m K. High value of thermal conductivity indicates that the material is good conductor. Low value of thermal conductivity indicates that the material is a good insulator and poor conductor [10]. The thermal conductivity is not always constant. It increases with the increase of ambient temperature, the moisture of material, and the density of a material.

The inner structure of a material highly affects thermal conductivity. For example, light porous materials with large proportions of void in the structure (gas or air bubbles, not large enough to carry heat by convection) have the lowest thermal conductivities and act as good insulators. Dense solid materials usually have high levels of thermal conductivities and act as good conductors. Therefore, pure crystals have the highest thermal conductivities. Thermal conductivities of some building materials are given in Table 7.2.

There are many ways to measure the thermal conductivity of material which are based on the hot wire, guarded hot plate, modified hot wire, etc. Each is usually limited with the type of material. There are two basic techniques of thermal conductivity measurement such as in steady-state and unsteady state.

- The steady-state technique measures thermal conductivity at thermal equilibrium, which takes a long time. In this way, a constant heat source has to be assured.
- The unsteady state techniques are a faster way of thermal conductivity measurement. The measurement takes place during the heating up process.

In a guarded hot plate, a solid sample of material is placed between two plates, "hot plate" and "cold plate." "Hot plate" is heated, and "cold plate" is cooled. The temperatures (T_1 and T_2) on each side of the sample are monitored until they reach steady state and are used in the calculation of thermal conductivity along with the thickness of the sample and heat input. The scheme of guarded hot place is in Fig. 7.4.

Thermal conductivity λ in steady state is given by formula:

$$\lambda = \frac{\dot{Q} \cdot L}{A(T_1 - T_2)} \quad [\text{W/m K}] \tag{7.23}$$

where $\dot{Q}$ is heat conduction rate passing through the sample area in unit time [W]

Table 7.2 Thermal conductivity of selected building materials (at about 20 °C)

Material	Thermal conductivity—λ (W/m K)
Air (dry and quiet)	0.023
Aluminum and light alloys	125–200
Brick	0.65–0.80
Cellular glass	0.035–0.06
Concrete	1.2–1.75
Copper	401
Expanded polystyrene	0.035–0.045
Glass	0.6–1.38
Granite	2.80
Lightweight concrete	0.11–0.25
Limestone	1.50
Mineral or glass wool	0.04–0.08
Particleboard	0.1–0.13
PUR rigid foam	0.02–0.04
Steel	43–58
Timber (pine)	0.14
Water	0.60

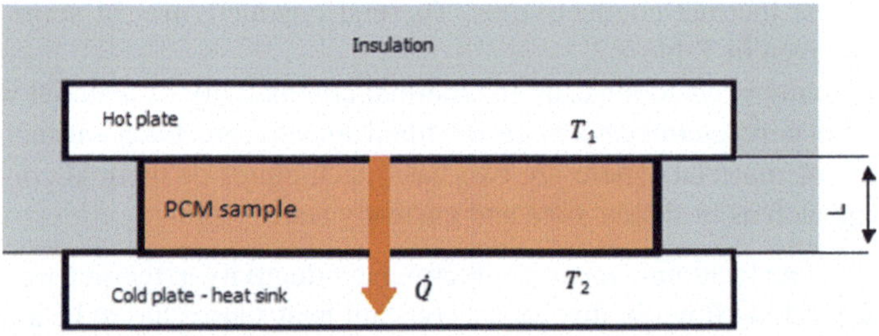

Fig. 7.4 Scheme of guarded hot plate

A area of the sample [m^2]
L thickness of the sample [m]
T_1 temperature on "hot side" of the sample [K]
T_2 temperature on the "cold side" of the sample [K].

7.3 Heat Convection

Convective heat transfer is the transfer of thermal energy between a solid surface and adjacent liquid/gas in motion. The nature of the process is highly dependent on the nature of the fluid motion adjacent to the surface where the boundary layer develops. Convective heat transfer to or from a solid requires the transfer of heat by the motion of the fluid and the transfer of heat by diffusion. The contribution due to random molecular motion dominates near the surface where the fluid velocity is low. At the interface between the surface and the fluid, the fluid velocity is zero (i.e., there is no relative motion between the fluid and surface) and heat transfer is transferred by diffusion only.

Convection is usually described as being either natural ("free") or forced. Natural convection occurs when the airflow is induced by buoyancy forces arising from the differences in density caused by temperature variations in the fluid in close proximity to the surface. Forced convection occurs when the flow over the transfer surface is assisted by fans, atmospheric winds, or pumps.

7.3.1 Newton's Law of Cooling

Newton's law of cooling is used to express the *rate of convection heat transfer,* which is proportional to the convective heat transfer coefficient h_{i}, surface area A, temperature difference between the surface T_{s} and surroundings T_{i}. The schematic representation of the heat convection by a wall is shown in Fig. 7.5, where h_{e} and h_{i} are exterior and interior wall surface heat transfer coefficients.

This can be described by the following expression

$$\dot{Q}_{\mathrm{conv}} = h_{\mathrm{i}} A (T_{\mathrm{s}} - T_{\mathrm{i}}) \quad [\mathrm{W}] \tag{7.24}$$

or per unit area

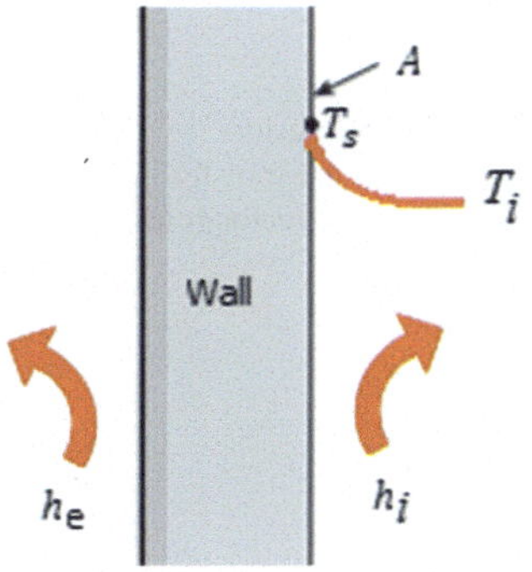

Fig. 7.5 Scheme of convection heat transfer

$$\dot{q}_{\mathrm{conv}} = h_{\mathrm{i}}(T_{\mathrm{s}} - T_{\mathrm{i}}) \quad [\mathrm{W/m^2}] \tag{7.25}$$

where

A surface area through which convective heat exchange takes place, m^2
T_s Surface temperature, °C
T_i interior air temperature far away from the surface, °C
h convective heat transfer coefficient of the process, W/m^2 K.

From the unit of convective heat transfer coefficient it can be observed that, convective heat transfer coefficient h represents a rate of heat transfer between solid surface and gas (or liquids) per unit area and temperature difference.

7.3.2 *Convective Heat Transfer Coefficient*

Convective heat transfer coefficient is not property of the fluid; it is experimentally determined parameter whose values depend on the:

- surface geometry,
- the type of media (gas or liquid)
- properties of the fluid in motion such as velocity, viscosity
- the nature of the motion, and
- temperature-dependent properties.

The value range for free and forced convection of h is given in Table 7.3. The values of convective heat transfer coefficient for rectangular buildings at exterior wall and roof as function of wind speed are given in Table 7.4.

Table 7.3 In general, the convective heat transfer coefficient for some common fluids

Fluid	Free convection (W/m^2 K)	Forced convection (W/m^2 K)
Air	5–25	10–200
Water	20–100	50–10,000

Table 7.4 Exterior convective heat transfer coefficient for rectangular building [11]

Wind speed (m/s)	Exterior wall 0°	Exterior wall 90°	Roof 0°
0.3	6	6	6
3	7.2	10	12
6	9	17	21
9	9.6	23	28.4
12	10	27.8	35.5
15	11.7	33	42

7.3.3 Conservation Laws in Differential Form

Heat transfer and fluid flow in the boundary layer are described by conservation equations of continuum mechanics, governing balance of mass, momentum, and energy.

7.3.3.1 Conservation of Mass

The conservation of mass states that the mass cannot be created or destroyed; thus, the mass within the control volume shown in Fig. 7.6 remains unchanged. Conservation of the mass is often called the continuity equation.

Therefore, the differential equation of the mass conservation can be expressed as the mass that enters the control volume is equal to the mass that leaves the control volume.

Mass flow rate is the product of the density, velocity, and area normal to the flow. If this is applied to the control volume in Fig. 7.6, then in it can be expressed as [12]:

$$\underbrace{\rho u(\mathrm{dy}\cdot\mathrm{dz})+\rho v(\mathrm{dx}\cdot\mathrm{dz})+\rho w(\mathrm{dx}\cdot\mathrm{dy})}_{\text{Rate of mass in flow to the control volume}}=\underbrace{\begin{aligned}&\rho\left(u+\frac{\partial u}{\partial x}\mathrm{dx}\right)(\mathrm{dy}\cdot\mathrm{dz})\\&+\rho\left(v+\frac{\partial v}{\partial y}\mathrm{dy}\right)(\mathrm{dx}\cdot\mathrm{dz})\\&+\rho\left(w+\frac{\partial w}{\partial z}\mathrm{dz}\right)(\mathrm{dx}\cdot\mathrm{dy})+\frac{\partial\rho}{\partial t}\mathrm{dx}\cdot\mathrm{dy}\cdot\mathrm{dz}\end{aligned}}_{\text{Rate of mass outflow from the control volume for transient state}} \quad (7.26)$$

where ρ is fluid density in kg/m^3, u is velocity in x-direction in m/s, and v is velocity in y-direction in m/s.

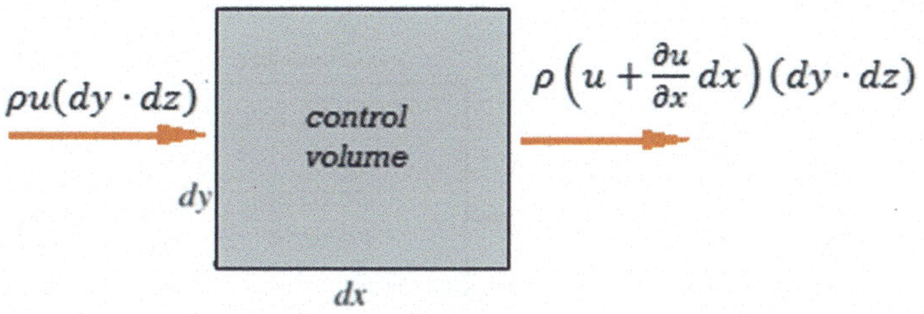

Fig. 7.6 Control volume from the boundary layer

Dividing Eq. (7.26) by $dx \cdot dy \cdot dz$ and subtracting the equation, a new simplified form is obtained as follows:

$$\frac{\partial \rho}{\partial t} + \rho \frac{\partial u}{\partial x} + \rho \frac{\partial v}{\partial y} + \rho \frac{\partial w}{\partial z} = 0 \tag{7.27}$$

7.3.3.2 Conservation of Momentum

The differential forms of the equations of motion in the velocity boundary layer are obtained by applying Newton's second law of motion to a differential control volume element (Fig. 7.7), and the differential form of the momentum equation can be obtained. The net force acting on the control volume represents the product of the mass and the acceleration of the fluid element within the control volume. Body forces and surface forces act on the control volume. Body forces act on the entire body of the control volume such as gravity, while surface forces act on the control surface such as the pressure forces. Therefore, the net body and surface force acting in a certain direction can be expressed as the product of the mass and the acceleration in that direction [12].

$$F_{\mathrm{body},x} + F_{\mathrm{surf},x} = \rho(\mathrm{d}x \cdot \mathrm{d}y \cdot 1) \cdot a_x \tag{7.28}$$

The surface forces acting in the x-direction become

$$\begin{aligned} F_{\mathrm{surf},x} &= \left(\tau + \frac{\partial \tau}{\partial y}\mathrm{d}y\right)\mathrm{d}x \cdot 1 - \tau \cdot \mathrm{d}x \cdot 1 + P \cdot \mathrm{d}y \cdot 1 - \left(P + \frac{\partial P}{\partial x}\mathrm{d}x\right)\mathrm{d}y \cdot 1 \\ &= \left(\frac{\partial \tau}{\partial y}\mathrm{d}y\right)\mathrm{d}x \cdot 1 - \left(\frac{\partial P}{\partial x}\mathrm{d}x\right)\mathrm{d}y \cdot 1 \\ &= \left(\frac{\partial \tau}{\partial y} - \frac{\partial P}{\partial x}\right)(\mathrm{d}x \cdot \mathrm{d}y \cdot 1) \end{aligned} \tag{7.29}$$

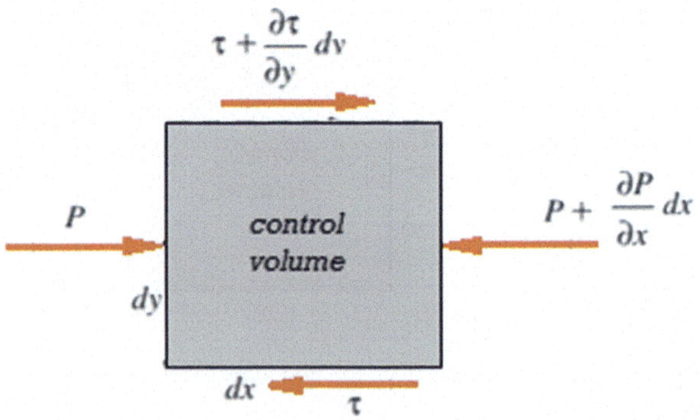

Fig. 7.7 Control volume from the boundary layer

Replacing $\tau = \mu \frac{\partial u}{\partial y}$ in Eq. (7.29), it is obtained

$$F_{\text{surf},x} = \left(\mu \frac{\partial^2 u}{\partial y^2} - \frac{\partial P}{\partial x}\right)(\mathrm{d}x \cdot \mathrm{d}y \cdot 1) \tag{7.30}$$

For two-dimensional steady-state flow, the velocity is

$$\mathrm{d}u = \frac{\partial u}{\partial x}\mathrm{d}x + \frac{\partial u}{\partial y}\mathrm{d}y \tag{7.31}$$

And acceleration is

$$a_x = \frac{\mathrm{d}u}{\mathrm{d}t} = \frac{\partial u}{\partial x}\frac{\mathrm{d}x}{\mathrm{d}t} + \frac{\partial u}{\partial y}\frac{\mathrm{d}y}{\mathrm{d}t} = u\frac{\partial u}{\partial x} + v\frac{\partial u}{\partial y} \tag{7.32}$$

Substituting all in Eq. (7.30) and dividing by $(\mathrm{d}x \cdot \mathrm{d}y \cdot 1)$, the final form is obtained

$$\rho\left(u\frac{\partial u}{\partial x} + v\frac{\partial u}{\partial y}\right) = \mu\frac{\partial^2 u}{\partial y^2} - \frac{\partial P}{\partial x}. \tag{7.33}$$

7.3.3.3 Conservation of Energy

The change of energy content of a system can be expressed as the difference between energy inflow and energy outflow.

$$\Delta E_{\text{sys}} = E_{\text{in}} - E_{\text{out}} \tag{7.34}$$

Applying this rule to a small control volume for a steady-state flow, the total energy content will remain constant; thus, no energy change of the system $(\Delta E_{\text{sys}} = 0)$.

Referring to Fig. 7.8, the rate of energy transfer to the control volume by mass in the x-direction is [12]:

$$\begin{aligned}(E_{\text{in}} - E_{\text{out}})_{\text{mass},x} &= \dot{m}c_{\text{p}}T - \left(\dot{m}c_{\text{p}}T + \frac{\partial\left(\dot{m}c_{\text{p}}T\right)}{\partial x}\mathrm{d}x\right) \\ &= -\frac{\partial\left(\dot{m}c_{\text{p}}T\right)}{\partial x}\mathrm{d}x\end{aligned} \tag{7.35}$$

where c_{p} is specific heat in J/kg K. Replacing $\dot{m} = \rho u(\mathrm{d}y \cdot 1)$ in the previous equation, it is obtained

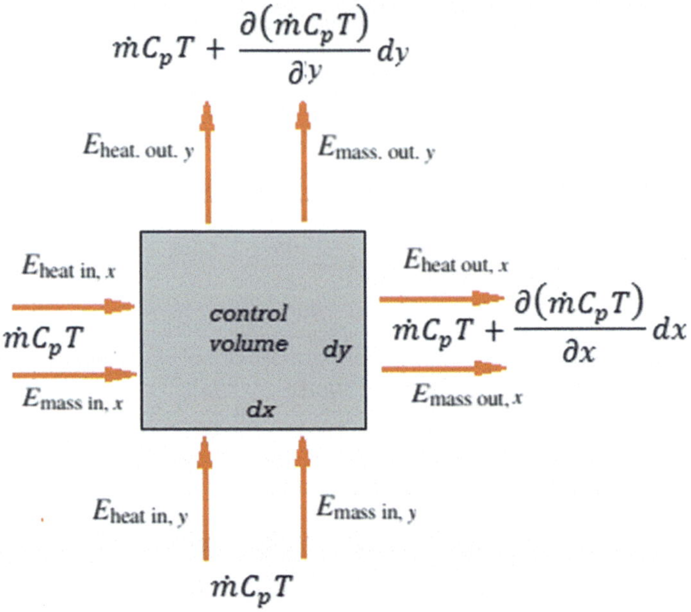

Fig. 7.8 Control volume for the energy transfers by heat and mass flow

$$(E_{\text{in}} - E_{\text{out}})_{\text{mass},x} = -\frac{\partial\left[\rho u\left(\text{d}y\cdot 1\right)c_{\text{p}}T\right]}{\partial x}\text{d}x = -\rho c_{\text{p}}\left(u\frac{\partial T}{\partial x} + T\frac{\partial u}{\partial x}\right)\text{d}x\text{d}y \tag{7.36}$$

Repeating the same procedure for any other direction and adding the results, the net rate of energy transfer to the control volume is obtained as follows:

$$\begin{aligned}(E_{\text{in}} - E_{\text{out}})_{\text{mass},y} &= -\rho c_{\text{p}}\left(u\frac{\partial T}{\partial x} + T\frac{\partial u}{\partial x}\right)\text{d}x\text{d}y - \rho c_{\text{p}}\left(v\frac{\partial T}{\partial y} + T\frac{\partial v}{\partial y}\right)\text{d}x\text{d}y \\ &= -\rho c_{\text{p}}\left(u\frac{\partial T}{\partial x} + v\frac{\partial T}{\partial y}\right)\text{d}x\text{d}y\end{aligned} \tag{7.37}$$

The net rate of energy transfer to the control volume by heat conduction in steady state is defined by Eq. (7.22). Therefore, combining Eqs. (7.22) and (7.37), the energy equation for the steady-state flow of a fluid with constant properties is defined as follows:

$$\rho C_{\text{p}}\left(u\frac{\partial T}{\partial x} + v\frac{\partial T}{\partial y}\right) = \lambda\left(\frac{\partial^2 T}{\partial x^2} + \frac{\partial^2 T}{\partial y^2}\right). \tag{7.38}$$

7.4 Heat Radiation

Heat transfer by thermal radiation does not require the presence of an intervening medium like convection and conduction. It takes place in the form of electromagnetic waves. All matter that has a temperature greater than absolute zero can emit electromagnetic waves. The higher the temperature of the matter, the higher the magnitude of the radiation. All solids, liquids, and gasses have the ability to absorb, emit, or transmit radiation. A surface that absorbs all incident radiation is called a *black surface* and emits energy at the maximum possible rate at a given temperature.

Stefan–Boltzmann law

The Stefan–Boltzmann law gives the maximum rate of radiation that is emitted from a surface [13]:

$$I = \varepsilon \sigma T^4; \quad \dot{Q}_{\mathrm{R}} = I \cdot A \tag{7.39}$$

where is

- I—solar irradiance ($\mathrm{W/m^2}$)
- $\dot{Q}_{\mathrm{R}}$—radiative heat transfer (W)
- σ—Stefan–Boltzmann constant ($\sigma = 5.67 \times 10^{-8}\ \mathrm{W/m^2\ K^4}$)
- T—temperature of the surface that emits radiation (K)
- A—area of the surface ($\mathrm{m^2}$)
- ε—radiative property of the surface called the emissivity ($0 \le \varepsilon \le 1$; for blackbody $\varepsilon = 1$, for gray body $\varepsilon < 1$).

Spectral radiation of the sun at the sea level and extraterrestrial radiation spectra is shown in Fig. 7.9.

Fig. 7.9 Spectral irradiance

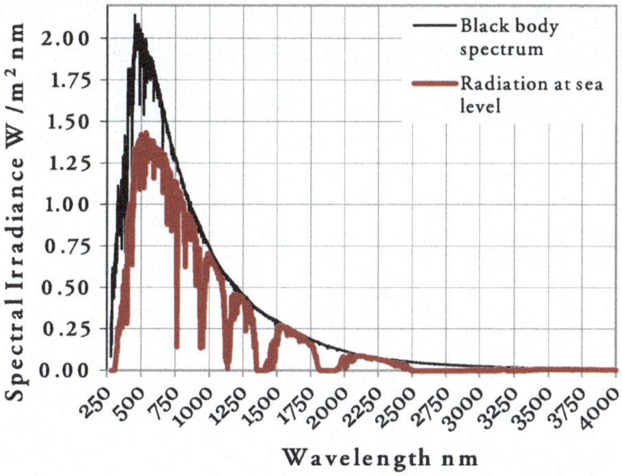

Nonblack surfaces do not absorb all incident radiation. When incident radiation reaches a surface, one part is absorbed; another is reflected or transmitted through the material as it is shown in Fig. 7.10.

Sum of all radiation components received by a glazed system is equal to incident radiation. In other words, sum of all fractions of incident radiation is equal to one.

Incident irradiation = absorbed + reflected + transmitted

$$I = aI + rI + \hat{\tau} I \tag{7.40}$$

$$a + r + \hat{\tau} = 1 \tag{7.41}$$

where

- a—absorptivity of the surface (fraction of incident radiant energy absorbed)
- r—reflectivity of the surface (fraction of incident radiant energy reflected)
- $\hat{\tau}$—transmissivity of the surface (fraction of incident radiant energy transmitted).

There are also special values in this case:

- For an opaque surface, $\hat{\tau} = 0$ and $a + r = 1$.
- For an absolute black surface, $a = 1$, ρ and $\hat{\tau} = 0$.
- For a gray surface $a = \varepsilon$.

If a surface area is completely enclosed by a much larger surface, e.g., a building under the sky, and separated by a gas such as air, the net heat exchange rate by the radiation can be calculated as

$$\dot{Q}_{\text{net}} = \dot{Q}_{\text{s}} - \dot{Q}_{\text{surr}} = \varepsilon\sigma A T_{\text{s}}^{4} - \alpha\sigma A T_{\text{surr}}^{4} \overbrace{=}^{\alpha=\varepsilon} \varepsilon\sigma A\left(T_{\text{s}}^{4} - T_{\text{surr}}^{4}\right) \tag{7.42}$$

where

- T_{s}—Temperature of the surface, (K)
- T_{surr}—Temperature of the surroundings, (K)

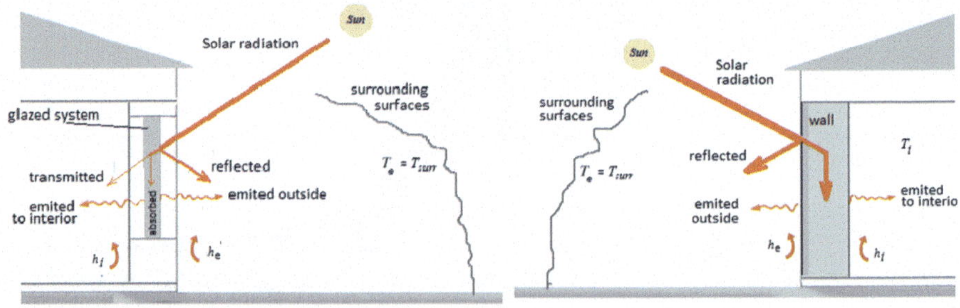

Fig. 7.10 Components of radiation

- $a = \varepsilon$ for the gray surface. If $T_s > T_{surr}$, the expression for $\dot{Q}_{net}$ is the same with the sign reversed, and $\dot{Q}_{net}$ is the net gain by A.

The previous formula is applicable only when the radiation is being considered. If there is a presence of convection between the surface and the gas, the contribution of the convection must be added to the previous equation. Note, convection is described by a driving potential based on the difference in the first power of the temperatures, whereas radiation is described by the difference in the fourth power of the temperatures, so it is useful to express net radiation as:

$$\dot{Q}_{net} = h_R A(T_s - T_{surr}) = \frac{T_s - T_{surr}}{\frac{1}{h_R A}}$$

$$h_R = \varepsilon\sigma\left(T_s^2 + T_{surr}^2\right)(T_s + T_{surr})$$

In this form, h_R depends on T_s which is often the desired result of the calculation. Thus, combined heat transfer composed of radiation and convection can be written as:

$$\dot{Q} = \dot{Q}_R + \dot{Q}_{conv} = h_R A(T_s - T_{surr}) + h_c A(T_s - T_{surr})$$

$$\dot{Q} = A(h_R + h_c)(T_s - T_{surr})$$

where

- h_R—radiation heat transfer coefficient
- h_c—convective heat transfer coefficient.

The temperature difference $T_s - T_{surr}$ is either Kelvin or °C, but the absolute temperatures must be used to calculate h_R. Coefficients h_R and h_c are always positive, but the heat transfer rate $\dot{Q}$ is determined by $(T_s - T_{surr})$ temperature difference sign.

7.5 Boundary Conditions

The boundary condition is obtained from energy balance at the surface. Therefore, heat energy transferred to the surface is equal to the heat energy transferred from the surface.

$$\dot{q}_{in} = \dot{q}_{out} \tag{7.43}$$

For conduction heat transfer mode, heat flux in positive x-direction can be expressed as

$$\dot{q}_{cond} = -\lambda \frac{\partial T(x,t)}{\partial x}$$

Note, the sign of heat flux is negative if it is in the opposite direction of the coordinate axis and vice versa.

For convection, the boundary condition is based on surface energy balance. Heat conduction at the surface is equal to the heat convection at the surface in a certain direction. This is the most common case in practice. For a simple heat transfer in one dimension through the wall, it can be expressed as

$$-\lambda\frac{\partial T(x,t)}{\partial x} = h(T_s(x,t) - T_\infty) \tag{7.44}$$

A combined boundary condition on a surface can be represented by the surface energy balance. Surface energy balance represents the heat transfer to the surface is equal to the heat transfer from the surface in all modes. For example, the boundary condition at the exterior wall surface of a building exposed to the natural environment consists of the balance of heat conduction at the surface that is equal to the sum of heat convection, thermal radiation, and solar irradiance in a certain direction as it is shown in Fig. 7.11 and expressed as

$$\dot{q}_{\text{in}} = aI - h_e[T_{es}(0,t) - T_e] - \varepsilon_e\sigma\left[T_{es}(0,t)^4 - T_e^4\right] = -\lambda\frac{\partial T(0,t)}{\partial x} \tag{7.45}$$

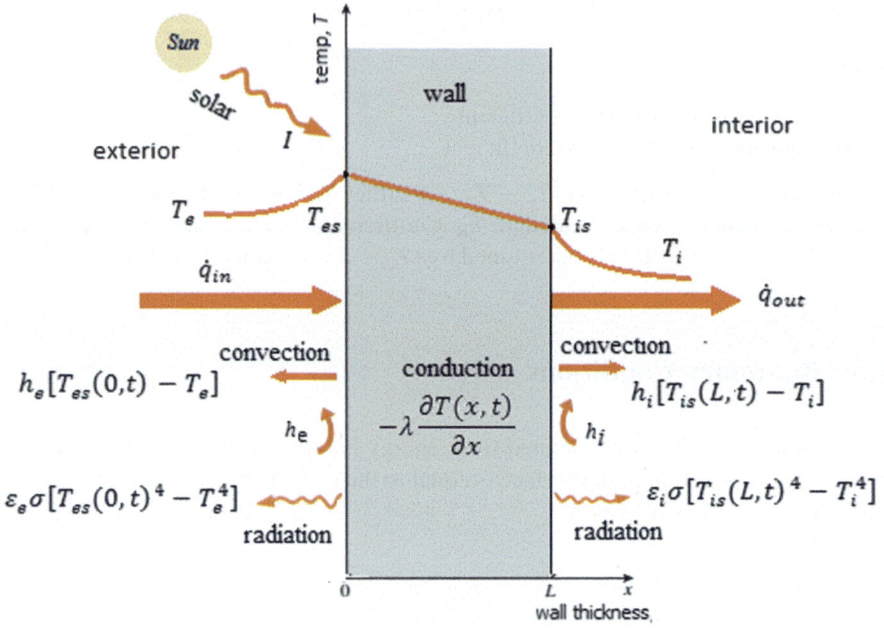

Fig. 7.11 Boundary condition at wall surface

where h_e is convective heat transfer coefficient in W/m^2 K, ε is emissivity, σ Stefan–Boltzmann constant is approximately 5.67×10^{-8} W/m^2 K^4, a absorptivity of the wall surface, I irradiance (incident solar radiation) in W/m^2.

Boundary condition at interior side of the wall can be expressed as the heat received by

$$\begin{aligned}\dot{q}_{\text{out}} &= h_i[T_{is}(L,t) - T_i] + \varepsilon_i\sigma\left[T_{is}(L,t)^4 - T_i^4\right] \\ &= -\lambda\frac{\partial T(L,t)}{\partial x}\end{aligned} \quad . \tag{7.46}$$

7.6 Thermal Resistance

Heat conduction through a wall expressed in Eq. (7.9) is divided by area A and rearranged as follows:

$$\dot{q}_{\text{cond}} = \frac{(T_1 - T_2)}{\frac{L}{\lambda}} \quad \left(\text{W/m}^2\right) \tag{7.47}$$

where $\frac{L}{\lambda}$ represents thermal resistance of a single-layer wall against heat conduction or just the *thermal conduction resistance* in $\frac{\text{m}^2\,\text{K}}{\text{W}}$.

$$R_{\text{cond}} = \frac{L}{\lambda} \quad \left(\text{m}^2\,\text{K/W}\right) \tag{7.48}$$

Figure 7.12 is a graphical explanation of thermal resistance due to conduction, convection, and radiation.

Thermal convection resistance

Heat convection from a wall expressed in Eq. (7.25) is rearranged as follows:

$$\dot{q}_{\text{conv}} = \frac{(T_s - T_\infty)}{\frac{1}{h}} \quad \left(\text{W/m}^2\right) \tag{7.49}$$

where $\frac{1}{h}$ represents thermal resistance of a square meter air layer next to the wall surface against convection or just the *convection resistance* in $\frac{\text{m}^2\,\text{K}}{\text{W}}$.

$$R_{\text{conv}} = \frac{1}{h} \quad \left(\text{m}^2\,\text{K/W}\right) \tag{7.50}$$

Figure 7.12 right shows graphically *thermal radiation resistance* between a wall surface at temperature T_s and the surroundings at temperature T_∞ and can be written as follows:

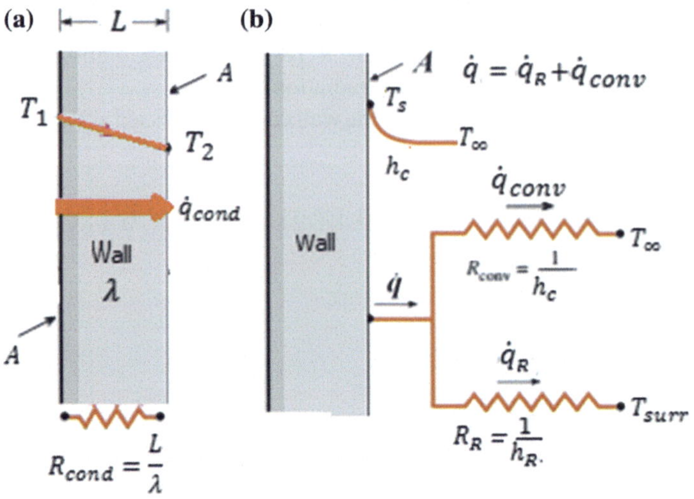

Fig. 7.12 Resistance due to conduction, convection, and radiation

$$\dot{q}_{\rm net} = \varepsilon\sigma\left(T_{\rm s}^4 - T_\infty^4\right) = h_{\rm R}(T_{\rm s} - T_\infty) = \frac{T_{\rm s} - T_\infty}{\frac{1}{h_{\rm R}}} = \frac{T_{\rm s} - T_\infty}{R_{\rm R}} \tag{7.51}$$

$$h_{\rm R} = \varepsilon\sigma\left(T_{\rm s}^2 + T_{\rm surr}^2\right)(T_{\rm s} + T_{\rm surr}) \tag{7.52}$$

The previous equation can valid only for the radiation component. In cases, the convection and radiation are present at a wall surface which is adjusted for convective component. Thus, combined heat transfer composed of radiation and convection can be written as:

$$\dot{Q} = \dot{Q}_{\rm R} + \dot{Q}_{\rm conv} = h_{\rm R}A(T_{\rm s} - T_{\rm surr}) + h_{\rm c}A(T_{\rm s} - T_{\rm surr})$$
$$\dot{Q} = A(h_{\rm R} + h_{\rm r})(T_{\rm s} - T_{\rm surr})$$

where

- $h_{\rm R}$—radiation heat transfer coefficient.
- $h_{\rm c}$—convective heat transfer coefficient (Fig. 7.13).

In this case, total resistance through the wall can be calculated by adding resistance on the exterior side of the wall with resistance due to conduction and total resistance for the parallel component on the interior side of the wall due to convection and radiation.

$$R_{\rm tot} = R_{\rm e,conv} + R_{\rm con} + \frac{R_{\rm i,conv}R_{\rm R}}{R_{\rm conv} + R_{\rm R}} \tag{7.53}$$

Thermal resistance of multilayer wall

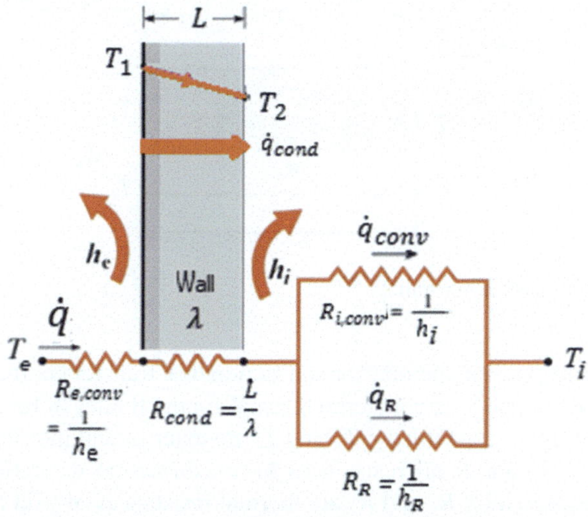

Fig. 7.13 Combination of series and parallel resistance (conduction, convection, and radiation)

Building walls are constructed of multiple layers. Usually, each layer has a different material that causes different thermal resistance. Consider the resistance of a two-layer wall that separates the exterior and the interior environment as it is shown in Fig. 7.14.

Heat flux for the wall is described as follows:

$$\dot{q} = \frac{(T_e - T_i)}{\frac{1}{h_e} + \frac{L_1}{\lambda_1} + \frac{L_2}{\lambda_2} + \frac{1}{h_i}} = \frac{(T_e - T_i)}{R_e + R_1 + R_2 + R_i} \quad \left(\frac{\mathrm{W}}{\mathrm{m}^2}\right) \tag{7.54}$$

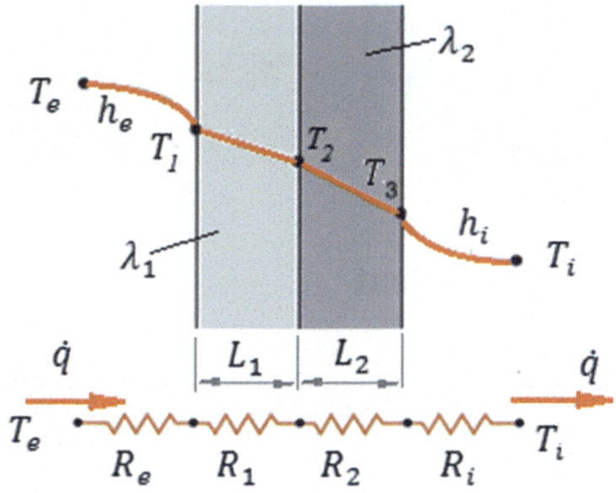

Fig. 7.14 Thermal resistance in series for two-layer wall

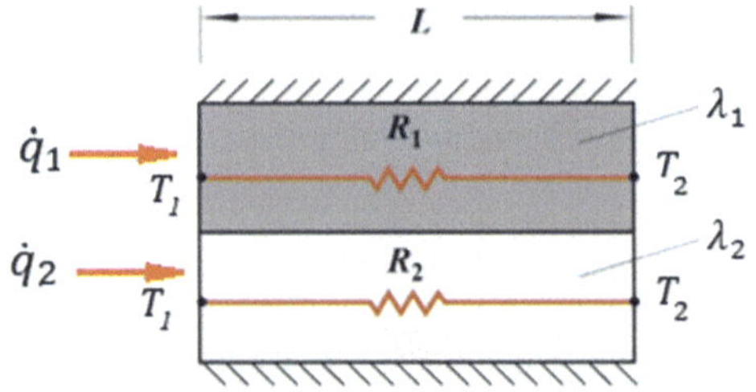

Fig. 7.15 Thermal resistance in parallel

where h_e and h_i are exterior and interior convection heat transfer coefficients, respectively, in W/m^2 K; λ_1 and λ_2 are material thermal conductivities of layer number one and layer number two, respectively; T_e and T_i are exterior and interior air temperatures, respectively; R_e and R_i are resistances due to convection at exterior and interior wall surfaces, respectively; R_1 and R_2 are thermal resistances of wall layers number one and number two, respectively.

$$R_{tot} = R_e + R_1 + R_2 + R_i = \frac{1}{h_e} + \frac{L_1}{\lambda_1} + \frac{L_2}{\lambda_2} + \frac{1}{h_i} \quad \left(\frac{m^2\,K}{W}\right) \tag{7.55}$$

Thermal resistance of two parallel layers shown in Fig. 7.15 can be defined from heat transfer equation for each layer separately.

$$\dot{q} = \dot{q}_1 + \dot{q}_2 = \frac{(T_1 - T_2)}{R_1} + \frac{(T_1 - T_2)}{R_2} = (T_1 - T_2)\left(\frac{1}{R_1} + \frac{1}{R_2}\right) \quad (W/m^2) \tag{7.56}$$

Or

$$\frac{1}{R_{tot}} = \frac{1}{R_1} + \frac{1}{R_2} \rightarrow R_{tot} = \frac{R_1 R_1}{R_1 + R_2} \tag{7.57}$$

7.7 Heat Transfer Coefficient

The heat transfer coefficient is the amount of heat, which is transferred in 1 s through 1 m^2 wall provided that temperature difference on both sides of the wall is 1 K, and heat flux is in steady state.

Overall heat transfer coefficient

$$U = \frac{1}{R_{tot}} = \frac{1}{\frac{1}{h_e} + \frac{L_1}{\lambda_1} + \frac{L_2}{\lambda_2} + \cdots + \frac{L_n}{\lambda_n} + \frac{1}{h_i}} \quad \left(\frac{W}{m^2\,K}\right) \tag{7.58}$$

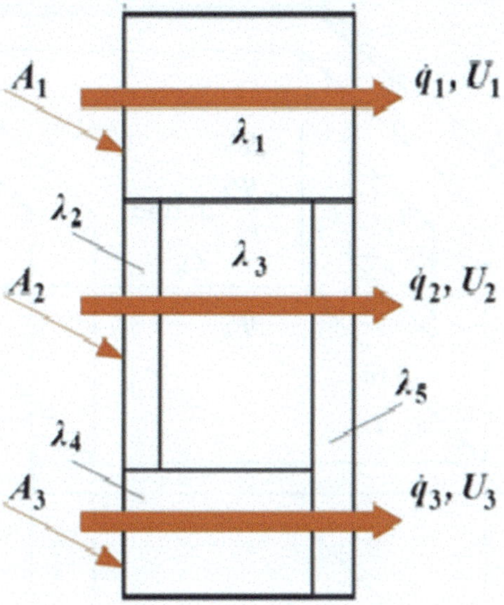

Fig. 7.16 Wall of simple heterogeneity

Overall heat transfer coefficient for a wall of simple heterogeneity (such as in Fig. 7.16a) is defined as follows:

$$U_{\mathrm{tot}} = \frac{\sum_{i=1}^{n} U_i A_i}{A_{\mathrm{tot}}} = \frac{U_1 A_1 + U_2 A_2 + \cdots + U_n A_n}{A_1 + A_2 + \cdots + A_n} \quad \left(\frac{\mathrm{W}}{\mathrm{m^2\,K}}\right) \tag{7.59}$$

Overall heat transfer coefficient for a wall of complex heterogeneity (such as in Fig. 7.16b) is defined as follows:

$$U_{\mathrm{tot}} = \frac{1}{A_{\mathrm{tot}}} \left(\sum_{i=1}^{n} U_i A_i + \sum_{j=1}^{m} \psi_j L_j + \sum_{k=1}^{p} \chi_k p \right) \quad \left(\frac{\mathrm{W}}{\mathrm{m^2\,K}}\right) \tag{7.60}$$

where ψ_j is linear heat transfer coefficient in $\left(\frac{\mathrm{W}}{\mathrm{m\,K}}\right)$, L is length relevant to ψ_j in m, χ_k is dotted heat transfer coefficient in $\left(\frac{\mathrm{W}}{\mathrm{K}}\right)$, and p is number of dotted connections (see Fig. 7.17).

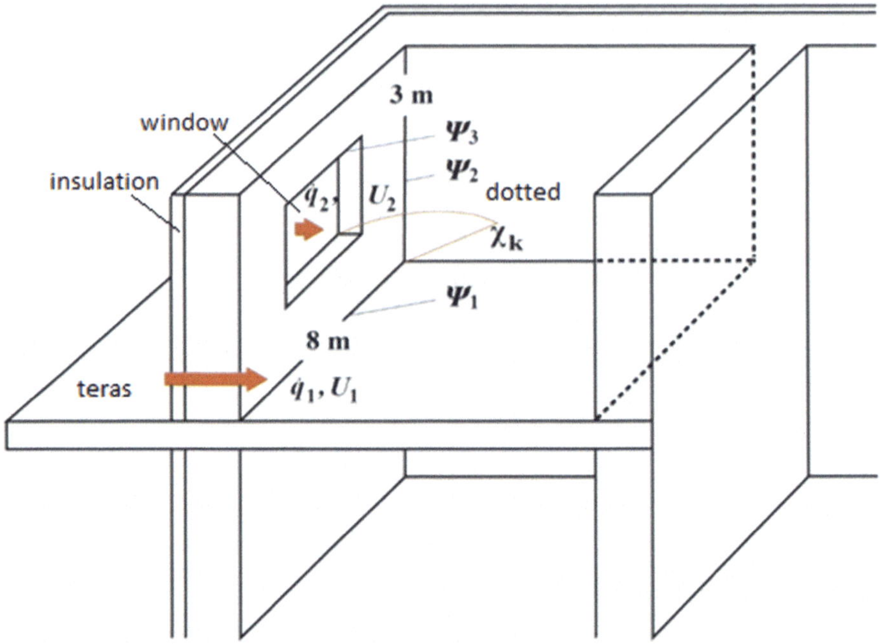

Fig. 7.17 Wall of complex heterogeneity (heat bridges)

7.8 Building Envelope System Modeling and Simulation Methods

Numerical research methods for the analysis of the process of the heat transfer through the building envelope and components as well as phase change are introduced. The most frequently used numerical techniques are the finite volume method (FVM) and the finite element method (FEM).

7.8.1 *Finite Volume Method*

The finite volume method (FVM) is a discretization method for approximation of the system of partial differential equations. A brief introduction of the mathematical model and finite volume discretization method is given in this subsection. Heat transfer and fluid flow are described by conservation equations of continuum mechanics, governing the balance of mass, momentum, and energy [14]:

$$\frac{\partial}{\partial t}\int_V \rho \mathrm{d}V + \oint_S \rho\vec{v}\mathrm{d}\vec{S} = 0 \tag{7.61}$$

$$\frac{\partial}{\partial t}\int_V \rho\vec{v}\mathrm{d}V + \oint_S \rho\vec{v}\otimes\vec{v}\mathrm{d}\vec{S} = -\oint_S p\mathrm{d}\vec{S} + \oint_S \tau\mathrm{d}\vec{S} \tag{7.62}$$

$$\frac{\partial}{\partial t}\int_V \rho h\mathrm{d}V + \oint_S \rho h\vec{v}\mathrm{d}\vec{S} = \oint_S \lambda\mathrm{grad}T\mathrm{d}\vec{S} + \int_V q\mathrm{d}V + \oint_S \vec{q}_S\mathrm{d}\vec{S} \tag{7.63}$$

The mass and momentum Eqs. (7.61) and (7.62) are solved in liquid-phase regions only, while the energy equation is solved both in the liquid and the solid phases of PCM.

In the equations given above, V is the control volume bounded by the closed control surface $\vec{S}$, t is the time, ρ is the density of the material, $\vec{v}$ is the velocity vector of the liquid phase of PCM, p is the pressure of the liquid phase of PCM, $\boldsymbol{\tau}$ is the shear stress tensor, h is the specific enthalpy, T is the temperature, λ is the thermal conductivity, q is the volume-based heat source or sink, and q_S is the surface-based heat source or sink.

Non-spherical part of the shear stress tensor $\boldsymbol{\tau}$ is related to the shear rate $\dot{\boldsymbol{\gamma}}$ using linear relation for Newtonian, incompressible, viscous fluids: $\boldsymbol{\tau} = 2\mu\dot{\boldsymbol{\gamma}}$, where μ is the dynamic viscosity of the liquid phase.

The specific enthalpy arising in Eq. (7.63) is defined as:

$$h = \int_{T_0}^{T} c_\mathrm{p}\mathrm{d}T + (1 - f_\mathrm{S})H \tag{7.64}$$

where c_p is the specific heat capacity of the PCM, H is its latent heat of phase change, and f_S is the volume fraction of the solid phase. In this work, the volume fraction variation with temperature is adopted to be linear from 1 to 0 between solidus and liquidus states.

The material properties, density, specific heat capacity, and thermal conductivity, are assumed to have constant, but different values depending on the phase of PCM.

Substituting h in Eq. (7.64) into Eq. (7.63), the latent heat contribution is added to the volume-based heat source term q. The surface-based heat source q_S may include the contribution of solar radiation in the cases where this is taken into account.

Resistance to the liquid-phase flow in the mushy region, caused by the solid fraction, is modeled analogously to the resistance of an isotropic porous medium. Following the Carman–Kozeny relation [15], the permeability K of the mushy region (in m^2) is given as a function of the solid volume fraction:

$$K = \frac{(1 - f_\mathrm{S})^3}{f_\mathrm{S}^2}\frac{1}{k\phi} \tag{7.65}$$

where k is a suitable model constant in m^{-2}, and ϕ is a smooth, non-dimensional switching function with the values from the range (0, 1], enabling de-/activation of the permeability model for low or high values of the solid volume fraction, respectively.

The integral conservation Eq. (7.63) applies to the solution domain as a whole, as well as to each cell in the grid. The equation discretization allows the use of unstructured grids consisting of arbitrary polyhedral, contiguous, non-overlapping cells and follows the finite volume approach by Demirdžić and Muzaferija [16] and Teskeredžić et al. [17].

The radiative heat transfer is calculated using surface-to-surface model. The net radiant flux on each surface in the domain is calculated demanding the equilibrium of radiative transfer over the entire closed set of surfaces, which depends on how each surface exchanges radiation with all other surfaces. This is accomplished by having the surfaces subdivided into patches and calculating the view factors for all patch pairs [18]. The irradiation I_{f} at each face f is equal to the irradiation on the associated patch I, while the face radiosity J_{f} is a summation of local face emitted energy E_{f} and the reflected irradiation $\tilde{\rho}_{\mathrm{f}} I_{\mathrm{f}}$ over individual spectra or bands:

$$I_{\mathrm{f}} = \sum I_{\mathrm{f,i}} \tag{7.66}$$

$$J_{\mathrm{f}} = \sum E_{\mathrm{f}} + \tilde{\rho}_{\mathrm{f}} I_{\mathrm{f,i}} \tag{7.67}$$

Radiative heat transfer equation and energy equation are linked on the glazing boundaries [19] via the surface-based heat source $q_{\mathrm{S}} = q_{\mathrm{b,rad}}$, so that the net radiative flux (absorption minus emission) is added to the exterior boundary heat flux caused by convection and conduction:

$$q_{\mathrm{b,tot}} = q_{\mathrm{b}} + q_{\mathrm{b,\,rad}} \tag{7.68}$$

$$q_{\mathrm{b,rad}} = \sum [(1 - \tau_{\mathrm{f}}) I_{\mathrm{f}} - J_{\mathrm{f}}] \tag{7.69}$$

where $\tilde{\tau}_{\mathrm{f}}$ is transmissivity of patch face, I_{f} irradiation at each patch face, and J_{f} is radiosity of the patch face and represents reflected and emitted radiation. When the incoming radiation component is to be neglected, for example, in the case of cloud weather, night time or north-facing glazing, $q_{\mathrm{b,rad}}$ is set to zero.

Boundary conditions on the exterior glass surface:

$$\dot{q}_{\mathrm{b,ex}} = \alpha_{\mathrm{ex}}(T_1 - T_{\mathrm{ex}}) + I_{\mathrm{n}} \tag{7.70}$$

Boundary conditions on the interior glass surface:

$$\dot{q}_{\mathrm{b,in}} = \alpha_{\mathrm{in}}(T_{\mathrm{in}} - T_4). \tag{7.71}$$

7.8.2 Finite Element Method

FEM is a numerical technique that describes a problem with partial differential equations. A domain is determined by a finite number of finite elements. The approximating function is determined in terms of nodal values. A continuous physical problem is discredited in a number of finite elements with unknown nodal values. The basic idea in the finite element analysis is to find a solution to a complex problem with simplifying the problem. The first step in FE analysis is to discretize the continuum into number of discretized finite elements. A general example for heat conduction is provided as follows [20]:

where

Element temperature

$$T = [N(x, y, z)] \times \{T\} \tag{7.72}$$

Interpolation Function

$$[N(x, y, z)] = \begin{Bmatrix} N_i & N_j & N_k & N_m \end{Bmatrix} \tag{7.73}$$

Nodal temperature vector

$$[T] = \begin{Bmatrix} T_i & T_j & T_k & T_m \end{Bmatrix}^{\mathrm{T}} \tag{7.74}$$

Temperature gradient in the element can be obtained by differentiation of the temperature interpolation equation:

$$\begin{Bmatrix} \frac{\partial T}{\partial x} \\ \frac{\partial T}{\partial y} \\ \frac{\partial T}{\partial z} \end{Bmatrix} = \begin{bmatrix} \frac{\partial N_i}{\partial x} & \frac{\partial N_j}{\partial x} & \frac{\partial N_k}{\partial x} & \frac{\partial N_m}{\partial x} \\ \frac{\partial N_i}{\partial y} & \frac{\partial N_j}{\partial y} & \frac{\partial N_k}{\partial y} & \frac{\partial N_m}{\partial x} \\ \frac{\partial N_i}{\partial z} & \frac{\partial N_j}{\partial z} & \frac{\partial N_k}{\partial z} & \frac{\partial N_m}{\partial x} \end{bmatrix} \{T\} = [B]\{T\} \tag{7.75}$$

The matrix $[B]$ is a temperature gradients' interpolation.

$$[B] = \begin{bmatrix} \frac{\partial N_i}{\partial x} & \frac{\partial N_j}{\partial x} & \frac{\partial N_k}{\partial x} & \frac{\partial N_m}{\partial x} \\ \frac{\partial N_i}{\partial y} & \frac{\partial N_j}{\partial y} & \frac{\partial N_k}{\partial y} & \frac{\partial N_m}{\partial x} \\ \frac{\partial N_i}{\partial z} & \frac{\partial N_j}{\partial z} & \frac{\partial N_k}{\partial z} & \frac{\partial N_m}{\partial x} \end{bmatrix} \tag{7.76}$$

Galerkin method [20] is used to derive the element equation and to write the heat transfer equation as follows:

$$\int_V \left(\frac{\partial}{\partial x}(\dot{q}_x) + \frac{\partial}{\partial y}(\dot{q}_y) + \frac{\partial}{\partial z}(\dot{q}_z) - \dot{e}_{\text{gen}} + \rho c_{\text{p}} \frac{\partial T}{\partial t} \right) N_{\text{i}} \mathrm{d}V = 0 \tag{7.77}$$

where $\dot{e}_{\text{gen}}$ is generated energy inside the element.

By applying the boundary conditions in the above equation, the element equation will be obtained as follows:

$$\int\limits_{V}\left(\frac{\partial}{\partial x}(\dot{q}_x)+\frac{\partial}{\partial y}(\dot{q}_y)+\frac{\partial}{\partial z}(\dot{q}_z)-\dot{e}_{\text{gen}}+\rho c_{\text{p}}\frac{\partial T}{\partial t}\right)N_{\text{i}}\text{d}V=0 \tag{7.78}$$

$$\begin{aligned}\int\limits_{V}\rho c_{\text{p}}\frac{\partial T}{\partial t}N_i\text{d}V-\int\limits_{V}\left[\begin{matrix}\frac{\partial N_i}{\partial x} & \frac{\partial N_i}{\partial y} & \frac{\partial N_i}{\partial z}\end{matrix}\right]\{\dot{q}\}\text{d}V\\ =\int\limits_{V}\dot{e}_{\text{gen}}N_i\text{d}V-\int\limits_{S}\{\dot{q}\}^{\text{T}}\{n\}N_i\text{d}S+\int\limits_{S}\dot{q}N_i\text{d}S-\int\limits_{S}h(T-T_{\text{e}})N_i\text{d}S\\ -\int\limits_{S}\left(\sigma\varepsilon T^4-a\dot{q}_{\text{R}}\right)N_i\text{d}S\end{aligned} \tag{7.79}$$

where $\{\dot{q}\}^{\text{T}}$ is heat flux across boundary, and $\{n\}^{\text{T}}$ is the direction cosine to outward normal.

$$\begin{aligned}\{\dot{q}\}^{\text{T}} &=\left[\begin{matrix}\dot{q}_x & \dot{q}_y & \dot{q}_z\end{matrix}\right]\\ \{n\}^{\text{T}} &=\left[\begin{matrix}n_x & n_y & n_z\end{matrix}\right]\end{aligned} \tag{7.80}$$

Note, it is

$$\{\dot{q}\}=-\lambda[B]\{T\} \tag{7.81}$$

The heat balance Eq. (7.79) can be broken down in the following discrete elements:

$$[C]\{\dot{T}\}+([K_{\text{c}}]+[K_{\text{h}}]+[K_{\text{R}}])\{T\}=\{R_{\dot{\text{e}}}\}+\{R_{\text{T}}\}+\{R_{\dot{\text{q}}}\}+\{R_{\text{h}}\}+\{R_{\text{R}}\} \tag{7.82}$$

where $[C]$ is heat capacitance matrix; $[K_{\text{c}}]$ is conductivity matrix; $[K_{\text{h}}]$ convective matrix; $\{R_{\dot{\text{e}}}\}$ is the heat generation matrix; $\{R_{\text{T}}\}$ is the heat flux across the boundary S_1; $\{R_{\dot{\text{q}}}\}$ is the heat flux across the boundary S_2; $\{R_{\text{h}}\}$ is the convective heat flux across the boundary S_3; $\{R_{\text{h}}\}$ is radiative heat flux across the boundary S_4.

$$[C]=\int\limits_{V}\rho c_{\text{p}}[N]^{\text{T}}[N]\text{d}V \tag{7.83}$$

$$[K_{\text{c}}]=\int\limits_{V}\lambda[B]^{\text{T}}[B]\text{d}V \tag{7.84}$$

$$[K_{\mathrm{h}}] = \int_{S_3} h[N]^{\mathrm{T}}[N]\mathrm{d}S \tag{7.85}$$

$$[K_{\mathrm{r}}]\{T\} = \int_{S_4} \sigma\varepsilon T^4[N]\mathrm{d}S \tag{7.86}$$

$$[R_{\dot{\mathrm{e}}}] = \int_{V} \dot{e}_{\mathrm{gen}}[N]^{\mathrm{T}}\mathrm{d}V \tag{7.87}$$

$$[R_{\mathrm{T}}] = -\int_{S_1} \{\dot{q}\}^{\mathrm{T}}\{n\}[N]^{\mathrm{T}}\mathrm{d}S \tag{7.88}$$

$$[R_{\dot{\mathrm{q}}}] = -\int_{S_2} \dot{q}_{\mathrm{s}}[N]^{\mathrm{T}}\mathrm{d}S \tag{7.89}$$

$$[R_{\mathrm{h}}] = -\int_{S_3} hT_{\mathrm{e}}[N]^{\mathrm{T}}\mathrm{d}S \tag{7.90}$$

$$[R_{\mathrm{R}}] = -\int_{S_4} a\dot{q}_{\mathrm{R}}[N]^{\mathrm{T}}\mathrm{d}S. \tag{7.91}$$

7.9 Selected Software Packages for PCM Simulation

There are different commercial and open-source simulation programs for full-scale building simulation and building components but only a few of them have the capability to analyze the phase change process. Some of them are TRNSYS, DOE-2, PCM Express, Energy 10, BLAST, EnergyPlus, TRNSYS, ESP-r, RadCool, IDA-ICE3, HVACSim, CLIM2000, WUFI, DÄMMWERK, etc. Numerical simulation tools that capable to model and simulate phase change in buildings are listed in Table 7.5.

StarCCM+, ANSYS Fluent, and MATLAB are not intentionally developed for building simulation, but they are very useful engineering tools that have integrated modules for phase transition simulation. The other simulation tools are intentionally developed for full-scale building simulation with the capability of simulating PCM along with the heat transfer.

Table 7.5 Some computer programs with capability to simulate PCM along with heat transfer tools

Computer program	References
TRNSYS	http://www.trnsys.com/
RadCool	
PCM Express	https://www.pcm-express.com/
EnergyPlus	https://energyplus.net/
ESP-r	http://www.esru.strath.ac.uk/Programs/ESP-r.htm
WUFI	https://wufi.de/en/
DOE-2	http://www.doe2.com/
StarCCM+	https://www.plm.automation.siemens.com/global/en/products/simcenter/STAR-CCM.html
ANSYS Fluent	https://www.ansys.com/products/fluids/ansys-fluent
MATLAB	https://www.mathworks.com/products/matlab.html
DÄMMWERK	https://www.bauphysik-software.de/de-de/

7.10 Remarks

To estimate the performance of passive design concept, the basics of mathematical models of physical processes were introduced. The models were developed based on numerical and experimental results. Three major areas of physical processes were discussed: heat transfer through the building envelope, heat flow through solar chimney, and energy storage within PCM. For solar chimney, optimum inclination angle is 45–60° for roof solar chimneys. Optimum openings size is required bigger with bigger height and cross section of the solar chimney. Environmental conditions such as wind speed and solar radiation as well as the orientation of the building are major influencing factors for passive design of buildings. Mathematical models for heat transfer through the building envelope and from the building envelope to the interior, as well as energy storage within PCM, are introduced in one place and are a good reference for the professionals, researches, academicians, and students.

References

1. Duraković B, Mešetović S (2019) Thermal performances of glazed energy storage systems with various storage materials: an experimental study. Sustain Cities Soc 45:422–430
2. Durakovic B, Torlak M (2017) Simulation and experimental validation of phase change material and water used as heat storage medium in window applications. J Mater Environ Sci 8(5):1837–1846
3. Durakovic B, Torlak M (2017) Experimental and numerical study of a PCM window model as a thermal energy storage unit. Int J Low-Carbon Technol 12(3):272–280
4. Afonso C, Oliveira A (2000) Solar chimneys: simulation and experiment. Energy Build 32(1):71–79

5. Shen J, Lassue S, Zalewski L, Huang D (2007) Numerical study on thermal behavior of classical or composite Trombe solar walls classical Trombe wall. Energy Build 39(8):962–974
6. Sakonidou E, Karapantsios T, Balouktsis A, Chassapis D (2008) Modeling of the optimum tilt of a solar chimney for maximum air flow. Sol Energy 82(1):80–94
7. Dimoudi A (2009) Solar chimneys in buildings—the state of the art. Adv Build Energy Res 3(1):21–44
8. Andersen K (1995) Theoretical considerations on natural ventilation by thermal buoyancy. ASHRAE Trans 15:81–93
9. Shi L, Zhang G, Shi L, Zhang GM (2016) An empirical model to predict the performance of typical solar chimneys considering both room and cavity configurations. Build Environ 103:250–261
10. Durakovic B, Yildiz G, Yahia ME (2020) Comparative performance evaluation of conventional and renewable thermal insulation materials used in building envelops. Tehnicki vjesnik - Tech Gaz 27(1) (in Press)
11. Hens HSL (2012) Applied building physics: boundary conditions, building performance and material properties front cover. Wiley, Darmstadt, Germany
12. White FM (2011) Fluid mechanics. McGraw Hill, New York
13. ASHRAE handbook of fundamentals (2009). American Society of Heating, Refrigerating and Air-Conditioning Engineers, Inc., USA
14. Ferziger JH, Peric M (1997) Computational methods for fluid dynamics. Springer, New York
15. Carman P (1937) Fluid flow through granular beds. Trans Inst Chem Eng 15:150–166
16. Demirdžić I, Muzaferija S (1995) Numerical method for coupled fluid flow, heat transfer and stress analysis using unstructured moving meshes with cells of arbitrary topology. Comput Methods Appl Mech Eng 125(1–4):235–255
17. Teskeredžić A, Demirdžić I, Muzaferija S (2002) Numerical method for heat transfer, fluid flow, and stress analysis in phase-change problems. Numer Heat Transf Part B 42(5):437–459
18. CD-Adapco (2013) User guide—STAR-CCM+ version 8.06. CD-Adapco
19. Viskanta R (1966) Radiation transfer and interaction of convection with radiation heat transfer. In: Advances in heat transfer. Academic Press Inc, New York, pp 175–252
20. Nikishkov GP (2004) Introduction to the finite element method. Lecture notes. University of Aizu, Aizu-Wakamatsu, Japan

Chapter 8
Conclusion

Phase change materials (PCMs) for thermal energy storage in buildings have been known for more than half of the century. The first experiment with PCM in building application is done by Maria Talkes, Assistant at MIT's Department of Metallurgy in 1948. At the time, it did not grab the attention of researchers, but sporadic study results were appeared in scientific publications until two decades ago. In the past fifteen years, application PCMs in buildings became a very active research area grabbing the attention of many researches over the world. A rapidly growing number of published papers are based on experimental and numerical studies. Most of the studies come from predominantly hot climate areas, placing China at the top of the list with a share of about 30%.

PCM-based technologies for building application have the potential to be integrated within walls, floors and ceilings, glazed systems, and air-based/free cooling systems. All these categories have different benefits, trends, issues, and opportunities, but they can still be one category as a part of new researches and experimentations. The challenges related to new technologies are to utilize opportunities, improve all benefits, and address the issues. Some issues are as high initial cost, loss of phase change capability, corrosiveness (in cases of some inorganic PCMs), and issues associated with the optimal design of containers have hampered widespread adoption. Paraffins generally perform well but due to their flammability property, they are able to catch fire easily; thus, more attention is paid to PCMs based on fatty acids or inorganic salt hydrates [1].

8.1 Trends

Fighting greenhouse gases became a global trend and driver for various technologies. Efforts of different regulatory bodies to control greenhouse gas emission resulted in deploying various environment-friendly technologies that reduce greenhouse emission and save energy. PCM-based technologies fit this scope and became a hot topic for the research with significant market growth. Recent trends are reflected in demand

© Springer Nature Switzerland AG 2020

B. Duraković, *PCM-Based Building Envelope Systems*, Green Energy and Technology, https://doi.org/10.1007/978-3-030-38335-0_8

for application of PCM in heating and cooling applications such as passive building heating and cooling systems, HVAC systems, textiles' cloth, packaging and electronic cooling. Using these materials in these applications will result in reducing energy demand for heating/cooling and consequently will reduce greenhouse gases. This global trend caused a significant expansion in the PCM market as well.

Energy saving and sustainable development became a global trend that is mostly reflected in the buildings and construction industry. PCM-based solutions play a major role in this game with the fastest growing regional market and research activity in Asia-Pacific. Based on the Global Market Insights, Inc. report, it is expected that the global market of PCM will reach a worth of 4 billion dollars by 2024.

But few things should be taken into consideration when choosing this material in any shape or form in order to improve the thermal comfort of the interior. As it has been recommended and concluded by several authors [2–5] that PCM should be carefully chosen since every sort of PCM reacts differently in contact with other materials due to its different origins. The life span of PCM should be investigated before implementing it in order to ensure financial feasibility and reduction in the cost of energy expenditure. The climatic region should be thoroughly investigated in order to decide whether PCM will make any impact if implemented since its melting–solidifying cycle needs to complete in a 24-h frame in order to be efficient and utilize its potential. The positioning of the PCM integrated element is of crucial importance for its performance since not all climatic environments are the same nor does PCM perform at its best in any given position. Lastly, the thicknesses of layers in the building envelope are of important significance since PCM can withhold a certain amount of energy. Other layers need to be in their optimal thickness in order to accommodate the ultimate energy absorption and release via the PCM medium. Current and future trends are reflected in the development of more efficient PCMs, more efficient design solutions of building components and modules, and looking for cheaper manufacturing process of PCMs.

8.1.1 Trends with Glazing Systems

The glazing system design requirements such as heating, cooling, cost, daylighting, and aesthetics are often in conflict. Since the energy-efficient design of a glazing depends as well on used materials and orientation, the design recommendation and decision criteria are considered below. Part of the overall thermal resistance of a building fenestration system derives from convective and radiative heat transfer between the exposed surfaces and the environment and in the cavity between glazing panes. Surface heat transfer coefficients h_e, h_i, at the exterior and interior glazing surfaces, respectively, combine the effects of radiation and convection. Wind speed and building orientation are important in determining the impact of h_e on the heat transfer.

The purpose of this is fulfilled by finding significant differences in overall performances among various materials applied in the responsive glazing systems studied in the literature recently. Also, impacts of different environmental conditions on the performance of various glazing systems have been studied. The relationship between the environmental conditions and design parameters such as heat gain and comfort time has been found. Therefore, the objectives of the research have been fulfilled as follows:

Experimental investigation of the glazing model

An experimental study on PCM-filled glazing samples exposed to the environment with and without radiation is studied. The results are compared to the conventional air-filled glazing. The experiments were conducted in a laboratory environment for convective–conductive tests. The natural environment is used for sample testing in the presence of radiation. The main purpose of the laboratory experiments and natural environment experiments is the validation of the numerical model results.

The PCM sample has less temperature fluctuation for about three degrees and is able to accumulate the heat from convective and radiative components in form of latent heat by keeping the lower interior surface temperature in the vicinity of the phase transition temperature. The result is delaying overheating of the interior for the period needed to melt the material. Once the PCM is completely melted, there is an instantaneous jump of heat gain mainly due to radiation. Melting and solidification periods can be extended by increasing the thickness of the cavity [6].

Computational models

The computational models for the glazing system have been validated using experimental results. Thermal energy storage system computation models exposed to convective–conductive heat transfer and radiative heat transfer are formulated separately and experimentally validated. It was shown very good matching between experimental and numerical results. The results have been assessed as correct and reliable for predicting of glazing system performance under different environmental conditions. The model is adjusted to the local climate parameters and used as a predictor of the glazing system performances under different conditions. The impact of the exterior convective heat transfer coefficient and the cavity thickness on the glazing performance such as heat gain, inactivity time of air conditioning system, and its relationships are investigated in which maximum layer thicknesses of 24 mm are recommended.

Charging/discharging and unload time analysis

It has been found that the linear dependences are associated with models exposed to the natural setup and with the presence of solar radiation. Nonlinear dependences are associated with the model exposed to the controlled environment and the presence of convective–conductive heat transfer only. Both of these models have high values of the coefficient of determination ($R^2 > 0.9$), which makes a good fit for functions.

In the case of the model with the presence of solar radiation, the phase transition temperature periods are longer in the solidification process. In addition to the convective component of the heat transfer, solar heat flux fosters the melting process making it shorter. The melting process represents the charging hours while the solidification process represents discharging hours. In the case of models with the presence of convective–conductive heat transfer, transition periods (melting and solidification) are almost the same. Longer phase transition periods are beneficial to the A/C load.

Unload hours per day growth with the increase of cavity width for all models. In the case of convective–conductive heat transfer, unload time per day is formulated using a nonlinear function. There is rapid nonlinear growth of the function slope up to the cavity of 16 mm, and then the slop is slightly getting reduced, which is caused by solid volume fraction. In case of the presence of the radiation component, the unload time per day is described by two hours. The slope of the PCM curve gets reduced in the vicinity of 24 mm cavity, which is caused by remaining liquid volume fraction. Remaining solid and liquid volume fractions of the PCM cause a reduction in heat storage trend and consequently reduce unload time.

Performances analysis for different external heat transfer coefficients

The impact of the exterior convective heat transfer coefficient on the glazing system performances is significant. The coefficient is changeable with the wind speed over the glass surface, which varies from less than 0.2 m/s for calm weather, free convection conditions, to over 30 m/s for storm conditions. Accordingly, three different values (15, 25, and 35 W/m^2K) of the coefficient have been studied. The relationship between the coefficient values and unload time over various cavity widths is described with mathematical equations.

It was found that the convective heat transfer coefficient has a significant impact on the remaining solid/liquid volume fraction during the melting/solidification process. The nonlinear relationship between the exterior convective coefficient and the remaining solid volume fraction exists. Depending on the exterior convective coefficient, it was found that the recommended cavity width varies from 16 to 21 mm for $h_e = 35\ W/m^2K$.

Recommended cavity width identification

Recommended cavity widths have been identified per different design criteria such as load/unload time, charging/discharging time, heat gain, average temperature, and temperature variation. Simulation results show that the increase of PCM cavity thickness increases the benefit to the A/C unload hours. The increase of the cavity thickness is beneficial but limited by solid–liquid phase volume fraction. Thus, no more than a 24 mm cavity is recommended for the PCM-filled glazing exposed to the radiation; otherwise, no more than a 16 mm cavity is recommended in the absence of the radiation [6].

8.2 Issues to Be Addressed

8.2.1 PCM in Building Structure Issues

Incorporating PCM in building structure increases thermal inertia of building and helps to reduce interior temperature variance providing thermal comfort. Combining PCMs with thermal insulation materials may provide better results [7]. There are still some issues associated with the current state of PCM's technologies that have to be addressed.

The phase transition temperature is an important parameter in selecting the right material for PCM applied in building envelope structure and components. The higher phase transition temperature is more suitable for hot climates while the lower phase transition temperature is more suitable for colder climates. The most efficient values of phase transition temperature for passive systems would be those from the thermal comfort range. It is difficult to meet these two requirements in areas with significant temperature shift from summer average temperature to winter average temperature. Simply the efficiency of selected material will be reduced in one of the seasons. This issue can be addressed by suitable design of building envelope structure and components, or application of different PCMs with different phase transition temperatures, which will be an important and intensive research area in the future.

The amount of the material and the layer thickness is not investigated enough. Some research shows more building energy reduction with the decrease of PCM layer thickness and the increase of surface up to certain dimensions for the same amount of PCM. Also, PCM with lower thermal conductivity in some instances might be desirable, for PCMs integrated into the building envelope and some building components. This may reduce heat transfer between the interior and exterior before the phase transition starts.

8.2.2 PCM in Glazed Systems Issues

The thermal and optical characteristics of the glazing usually determine its applications. The thermal characteristics are the insulation including the type of the material in the cavity and its conductivities, overall glazing area, and the amount of framing and its material. The optical characteristics of the glazing system are defined by the material of the glass and cavity content.

The current state of the commercially available PCM does not provide good optical properties. Due to the opacity of the PCM in solid state, there are some limitations of the PCM-filled glazing systems application to some commercial and industrial buildings, which are significant energy consumers. This is an issue that has to be addressed in future work. Thus, it has limited application to the buildings with a lower requirement for the light.

Future work may be focused on high-performance glazing that can be efficiently used in summer and winter in combination with self-natural ventilation strategies. Basically, two main research directions on this subject at this time are essential including the research of new more efficient PCMs and design research.

Material research may include

- Higher enthalpy.
- Phase transition range 21 °C to maintain both winter and summer modes.
- Improve optical performances to be more suitable for glazing systems.

Design research

- Look for optimum design suitable to operate interchangeably in summer and winter modes.
- Continue numerical research for analysis of the impact different environmental factors on the design, using existing commercially available PCMs.
- Design testing in real condition in building and mechanical design issues.

Mechanical design issues that have to be addressed include opening and closing the flaps in summer and winter mode, weight issue, and condensation that may appear in the exterior cavity.

The flaps may be operated from the interior via a handle connected to the flaps through a mechanism allowing the users switching them from summer to winter mode and vice versa. This feature may add cost to the system as well. Once the technology is defined completely and finds application in the market, the cost of this feature in mass production will be negligible.

The application of the PCM in glazing systems adds some extra weight to the system (between 16 and 21 kg/m^2). This requires stronger (and heavier) glazing frames as well. The weight issue may be reduced using modular window wings.

8.2.3 PCM in Separate Heat Storage Module Issues

Latent heat thermal energy storage systems (LHTESs) in modules have the potential to decrease electrical energy demand on space cooling by shifting the on-peak load to off-peak load, reduce temperature variance and provide thermal comfort. In this case, LHTES acts as a cold storage unit during off-peak hours (low electricity rate hours) and during on-peak load the LHTES discharges by providing cooled air to living space.

Some issues that have to be addressed in modular LHTES systems are associated with incomplete solidification of liquid phase during the night, a very limited contact area between PCM and the air and inability to use a significant portion of PCM due to low thermal conductivity of PCM and due to low convective heat transfer coefficient. Some proposed solutions are to use forced ventilation in case of increasing convective heat transfer coefficient and heat module design optimization. Different enhancement techniques were introduced including shape redesign of the PCM module by adding fins, spheres or cylinders. For example, free cooling modular systems are suitable for climates with daily temperatures less than 15 °C.

The high cost of the manufacturing process may hinder market growth. Material research is based on developing a high-performance PCM with low cost. This is a major enabler for the widespread application of the PCMs for energy storage. The issues are associated to achieve this goal, such as low thermal conductivity of PCMs, which reduces the efficiency of the PCM module. PCMs with higher thermal conductivities are desirable for PCMs integrated into modules. This way it will enhance the heat transfer rate through the PCM and consequently enhance charging and discharging of the LHTES module which maximizes the efficiency of the module. Various enhancement techniques have been introduced such as application of metallic foams, application of nanoparticles with high thermal conductivity, expanded graphite, and encapsulations.

Metallic foams perform better than nanoparticle in terms of thermal conductivity enhancement since they have no problems associated with nanoparticles. Nanoparticles have a significant impact on conductivity, but non-uniform dispersion and collection on a certain place are the issues associated. Thermal conductivity enhancement based on expanded graphite depends on packaging density, surface area, mass fraction, and thickness of the expanded graphite. PCM encapsulation is very promising for interior and exterior applications due to direct contact with the environment and increased thermal conductivity.

Issues discussed here represent obstacles for more efficient PCM systems in building application. Solving these will lead to better thermal performances, which will become a significant enabler for market growth. Different methods, better materials and techniques, and lower cost of the manufacturing process are major areas for research, development, and innovation.

References

1. Kissock J, Kelly J, Hannig M, Thomas I (1998) Testing and simulation of phase change wallboard for thermal storage in buildings. In: Proceedings of international solar energy conference. Albuquerque, New Mexico, 14–17 June 1998
2. Faraji M (2017) Numerical study of the thermal behavior of a novel composite PCM/concrete wall. Energy Procedia 139:105–110
3. Ahangari M, Maerefat M (2018) An innovative PCM system for thermal comfort improvement and energy demand reduction in building under different climate conditions. Sustain Cities Soc 44:120–129
4. Hanchi N, Hamza H, Lahjomri J, Oubarra A (2017) Thermal behavior in dynamic regime of a multilayer roof provided with two phase change materials in the case of a local conditioned. Energy Procedia 139:92–97
5. Duraković B, Torlak M (2017) Simulation and experimental validation of phase change material and water used as heat storage medium in window applications. J Mater Environ Sci VIII(5):1837–1846. ISSN: 2028-2508. Copyright © 2017, University of Mohammed Premier Oujda Morocco

6. Duraković B, Torlak M (2017) Experimental and numerical study of a PCM window model as a thermal energy storage unit. Int J Low-Carbon Technol XII(3):272–280
7. Durakovic B, Yildiz G, Yahia ME (2020) Comparative performance evaluation of conventional and renewable thermal insulation materials used in building envelops. Tehnicki vjesnik—Technical Gazette 27(1) (in press)

The manufacturer's authorised representative in the EU is Springer Nature Customer Service Centre GmbH, Europaplatz 3, 69115 Heidelberg, Germany. If you have any concerns regarding our products, please contact ProductSafety@springernature.com

Printed and bound by CPI Group (UK) Ltd, Croydon, CR0 4YY
30/09/2026
02225458-0005